Globalization, Technology, and Philosophy

Globalization, Technology, and Philosophy

Edited by

David Tabachnick
and
Toivo Koivukoski

State University of New York Press

Published by
State University of New York Press, Albany

© 2004 State University of New York

For information, address State University of New York Press,
90 State Street, Suite 700, Albany, NY 12207

Production by Kelli Williams
Marketing by Michael Campochiaro

Library of Congress Cataloging-in-Publication Data

Globalization, technology, and philosophy / edited by David Tabachnick
and Toivo Koivukoski.
    p. cm.
    Includes bibliographical references and index.
    ISBN 0-7914-6059-2 (alk. paper) — ISBN 0-7914-6060-6 (pbk. : alk.
paper)
    1. International economic relations. 2. Globalization. 3. Technological
innovations—Economic aspects. 4. Social sciences—Philosophy.
I. Tabachnick, David. II. Koivukoski, Toivo.

HF1359.G598 2004
337—dc22   2003070445

10 9 8 7 6 5 4 3 2 1

Contents

Introduction

David Tabachnick and Toivo Koivukoski

> "We can hold in our minds the enormous benefits of technological society, but we cannot so easily hold the way it may have deprived us, because technique is ourselves."
>
> —George Grant, "A Platitude"

What is globalization? What is technology? We cannot fully understand these phenomena by accounting for their many manifestations, by listing the impacts of globalization or different technologies. Globalization is not simply world-wide markets and technology is not simply a set of neutral tools. They are expressions of our will to master our planet. To understand these related phenomena we must accept that something essential is at stake in them, something that changes the way we understand community and that touches us directly as human beings.

The authors in this collection make an effort to understand globalization and technology through the lens of philosophy. Conventional wisdom would have us believe that others are better suited to explain globalization and technology: economists, heads of state, bureaucrats, engineers, computer programmers, biochemists, or other technical experts. Philosophy, it might be argued, offers very little in the way of practical responses to the multiple challenges of the future. For those who would say this, philosophy is an interesting, albeit useless, academic subject.

1

Philosophers have long recognized this criticism. Consider the amusing story about the philosopher Thales that Aristotle recounts in Book I of the *Politics* [1258b15–1259a36]. As the story goes, Thales is reproached for living in poverty because he spent his whole life engaged in 'useless' philosophy. To prove his critics wrong, he used his observations of the stars to predict a bumper crop of olives, bought up all the olive presses at a low price, and later rented them out at a profit. This proves, Aristotle writes, "that it is easy for philosophers to become rich if they so desire, though it is not the business which they are really about." Philosophy is not to be judged based upon its usefulness—its ability to solve particular problems, or in this case to make money—but based upon its capacity to understand and explain the whole, hard as this may be. For us, this means understanding globalization and technology. Fortunately, the authors of the following essays have taken the time to do just this.

In the opening essay of the collection, W. R. Newell argues that technology and a new global postmodernist paradigm are "slowly corroding" the character of political community and disintegrating civic virtue and obligation, so much so that democratic civilization as we know it is threatened with extinction. He argues we are now experiencing a renewal of the tension between our yearnings for a sense of community and individual rights, and suggests that, far from being a place of stability and boredom, a globalized world will be unsteady and incendiary. Newell's concern extends to a description of a planetary technological transformation that does not simply include the rise of new global political and economic regimes but also a new, potentially illiberal conception of human being.

Darin Barney's essay takes a specific look at the affect of the Internet and digital technology on community. He argues that on-line virtual community is deprived of the central tenet of liberal politics: moral obligation. The relationship between virtual and real community may even be antagonistic, since the growth of digital communication contributes to the decay of real community and civil life. As in Newell's piece, this discussion leads to a central dilemma for contemporary peoples and nations: the acceleration of individual autonomy versus a basic human need for association with others. All of the good things about overcoming divisions of geography and social standing within the virtual sphere also allow an anonymous entrance and exit from relationships. Dissatisfactions are no longer met with calls for political, legislative or social reform but with a simple click of the mouse, that severs all ties and obligations. The problem, Barney argues, is that we have mistaken communication for community.

Bernardo Attias remarks that "left-leaning rhetorics seem to be turning up in the strangest places." He shows how the information revolution has co-opted the language of revolutionary politics, such that we may no longer be able to speak about pathways to alternative communities. This is an important theme of the book: our attempts at dissent are inculcated by technology and globalization.

In the same vein, Tom Darby argues that the old categories and metaphors that we used to understand our world—like Left and Right—no longer work. The disorientation that results is not an uncommon occurrence in the history of civilizations, but our crisis of understanding is unique in that our world—the sphere of our knowing and making—has no limits. Our world, which is the world of technology, is self-referential, relatively autonomous, progressively sovereign, and tends toward the systemization of nature both human and non-human. Thus defined, there is nothing outside of technology against which it could be judged. Rather, technology puts forward its own standard: efficiency. For Darby, this is the basis of the new planetary justice.

Don Ihde challenges many of the views put forward in these first essays. He asks "Which kind of globalization do we want and how do we go about getting it?" He argues that as technology shapes our planet we must become aware of its unpredictable consequences. For this reason, Ihde critiques both utopian and dystopian visions of globalization as unlikely if not ridiculous. Rather than either demons that must be exorcised or the saviours for our social ills, technology and globalization are processes that need to be managed through a new kind of civil involvement.

Andrew Feenberg's essay is a bridge between Parts one and two: community and humanity. Like Ihde, he argues that a new politics directed towards democratization can arise from within a technological order, but again, this requires that we set aside both dystopian and utopian visions of technology. Both are visions of technology from the outside, either as destructive to our humanity or as a guarantor of our happiness and freedom. We do not stand outside of technology, but this does not mean that we are committed to a rationalized social order directed only by efficiency. Resistances "inevitably arise" out of the limitations of technological systems, and motivated by a search for meaning, these resistances can affect the "future design and configuration" of our world. These resistances form the basis for a new technological politics and a new technological human being.

Whereas Part I examines the changes that technology and globalization affect upon our communities, the essays in Part II ask, "By what standard do we judge or even notice these changes? Does something of our humanity stand outside of technology and globalization?" These essays all give differing accounts of the status of the self within technology and globalization, and of the role of philosophy in the project of self-knowledge.

As a general introduction to the philosophy of technology, Arthur Melzer's essay is excellent. When his overview is coupled with his critique of the common approaches to technology, the urgency of the subject becomes apparent. He argues that the more we rail against technology, the more firmly we are held in its grip. Using examples from the Right, Left, and Center, he explains that critiques of technology are themselves technological. Realizing this, we must go behind these critiques and back to classical philosophy.

Trish Glazebrook's essay is an attempt to amend the silence of philosophers of technology on the topic of globalization. She calls upon Heidegger's teachings and extends them to ethical, political, and cross-cultural practices, showing how the logic of domination and control does not stop with the "things" of non-human nature, but includes human beings themselves.

Gilbert Germain puts forward that in threatening our given worldliness—our particular, spatial limits and our relation to objects not of our own making—technology and globalization threaten our humanity. Not only does this tendency remove the external limits that define our being, but as the outside world is brought within our immediate grasp, we cease to see technology as a mediating term: we disappear into our technology, our technology disappears into us, and both collapse into a world that we no longer see as external to ourselves.

Criticizing and reforming technology is no easy matter of recalling a humanist standard against which it can be judged. Ian Angus argues that this is so because the separation between the technical and the ethical upon which humanist evaluation rests is undercut by technology. The modern self sees the good as that which is within its power to procure, and according to this definition, the technical and the ethical are interwoven. To assert a truly humanist creed one must first understand human beings as limited beings within a given context. For us, this means understanding technology, since technology supplies the context for modern existence.

Horst Hutter calls upon Nietzsche as the thinker who most fully thought through the ambiguities and contradictions that define our technological age. Perhaps owing to this inheritance, the essay is jarring. Hutter writes that to master technology, we must first master ourselves; this means going behind the unity of the self to see what it masks—a multiplicity of warring powers—and going forward toward the creation of a new human being.

According to Charlotte Thomas, philosophy is necessary for an adequate understanding of technology, but technology undercuts the basis for philosophical thought. In a world measured by efficiency and usefulness, philosophy seems to have no place. While she voices some hope, she sees the public currency of philosophy being devalued as we are ever more directed by the necessary and impressed by the specialist.

The book ends with a short essay by Donald Phillip Verene. For all of the talk of the self and the value of the individual, Verene argues that as functional members of technological society we are cut off from the possibility of self-knowledge. For us, the self is essentially undetermined and has a hollow core: there is nothing to know of the self, only an empty drive to mastery, and an empty standard of truth as certainty.

One of the cautions raised by many of our authors is that philosophical questions about globalization and technology are not only rare but also threatened. Philosophical thinking about the whole is crowded out to make way for specialized, instrumental rationality. Our thinking has become a tool directed toward solving the problems of the world. As a consequence, most studies of globalization and technology deal with specific problems concerning global society, economics, the environment, etc. This books aims to do something different: to understand what globalization and technology are in terms of how they affect our communities and our humanity. Though this may not directly solve the "problems" of technology or globalization, the openness to the whole that inspires these kinds of questions—the same wonder that caused Thales to contemplate the patterned changes in the heavens—may serve as a moderating influence on our mastery of the planet and ourselves, a program that would otherwise have only technological limits.

Part One

❖ ❖ ❖

Community

I

Democracy in the Age of Globalization

Waller R. Newell

Throughout history, the human soul has always expressed its longings for freedom and its capacities for virtue and vice through a particular ordering of the political and social community. For the ancient Greeks, it was the small cohesive city state or *polis*. For medieval Europe, it was the respective claims of pope and emperor. For the last two hundred years, it has been liberal democracy. First promulgated as an ideal during the Enlightenment, actualized with varying degrees of success during the nineteenth and twentieth centuries, until recently it formed the spiritual core of Western civilization. As we begin the new millennium, however, it looks increasingly as if this civilization may be coming to an end. Everywhere the Enlightenment project is in retreat or disrepute. More dangerously still for its survival, for many, the secular state with its representative institutions and procedural universalism has simply become boring. Liberal democracy often no longer engages people's primary loyalties, passions, or interests, which they are more and more likely to identify with their ethnic groups, issue groups, and a plethora of subcultures in which erotic and aesthetic proclivities can be freely indulged. Hence, while the non-Western world embraces its own premodern religious and cultural roots with renewed fervor and rejects the claim of liberal democracy to embody the single, universally valid path to the future, and while a host of demographic and economic catastrophes press in upon the liberal democratic heartland of Europe and North America, a spiritual malaise of ennui and disaffection eats away at the Western ethos from within. In many ways, we are standing blindfolded on the precipice of an enormous political, cultural, and economic upheaval comparable to the fall of the Roman Empire. At the outset of the millennium, it is entirely conceivable that liberal democracy is doomed.

The sources of its doom are ripening in the form of a dual assault on democratic civilization from the Right and the Left. On the one hand, we face the relentless dynamism of global technology and its impatience for the inherited customs, bonds, and institutions of the nation-state (exemplified by management guru Peter Drucker's call for the "reinvention" of the American political system to correct what he sees as the flaws in its economic efficiency stretching back to Locke and the Founding Fathers, those inconvenient political and civil institutions that have retarded our total transformation into producers and consumers of commodities and nothing else).[1] This is the continuation of what Marx regarded as the revolutionary mission of the bourgeoisie, the most radical revolution in history. Now worshipped as the global economic paradigm, it continues to uproot and destroy whatever may remain of vestigial human loyalties and bondedness. Hence, so conspicuous a success both as a financier and a citizen as George Soros has recently warned that capitalism is in danger of severing its links with the virtues of character previously thought to be the common source of civil society and commercial prosperity.[2]

On the other hand, we witness the continuing unfolding of the postmodernist agenda—the fragmentation of the nation-state into a kaleidoscope of ethnic and cultural tribalisms, self-invented "communities" and client groups comprised of a single, narrow biological or ideological fixation that detracts from any sense of shared civic obligations stretching across our substantive duties as citizens and family members. More perplexingly still, and contrary to the conventional wisdom, the dynamics of economic globalization are converging with the dynamics of postmodernism. Far from being opposed to one another, postmodernist deconstructionism and the global economic paradigm are actually cooperating and reinforcing each other in ways that are detrimental to civil society—a bizarre alliance in which Bill Gates joins hands with Jacques Derrida to deconstruct every inherited relationship and established usage. Although one side does this to remove the few constraints which the nation-state still imposes on economic globalization, while the other does so in order to replace these same constitutional and civic institutions with the primordial communities of gender and race, they cooperate to usher in a single goal—the disintegration of the nation-state into a multitude of idiosyncratic, self-absorbed tribalisms pursuing their illusory freedom within the gridlock of global technology.

We need to rethink the liberal tradition, including the bases of democratic civilization, civic virtue, and constitutional government, in light of the profound social, economic, and cultural transformations unfolding in the world today. In order to disentangle from these forces (what will preserve and nurture democratic civilization in contrast with what is harmful to it) we need to rethink the origins and character of modernity from the ground up. The place to begin is to reopen the debate over the meaning of history. Since the summer of 1989, it has been argued that we have reached the Hegelian "end of history." According to this argument, only the liberal democratic paradigm remains—actualized with uneven success so far outside of North America and Europe, but bound to prevail now that Marxism-Leninism, the last serious contender as a paradigm for legitimacy, has departed the historical stage. But in the years since Francis Fukuyama's formulation captured the public imagination, we have had ample reason to wonder whether any of this is really so. I would argue that the tensions Hegel diagnosed in 1806 continue in different forms, now that the particular variant of those tensions embodied in America's long struggle with the Soviet empire has passed from the scene.[3]

For, despite the collapse of Marxism-Leninism, dissatisfaction with the liberal democratic route to modernity—indeed, with the whole ethos of the Enlightenment—is arguably increasing, rather than decreasing. This dissatisfaction, manifested in a number of postmodernist social movements, is still rooted in the Rousseauian protest against modernity from which Marxism itself originally issued. Borrowing from Hegel, I call this ongoing revolution against liberalism the revolution of Understanding and Love. It underlies Marxism and it underlies the global and economic revolutions emerging in the postcommunist era. In order to grasp the forces behind this revolution, we must look again at Hegel. But it is a very different Hegel from the one identified by Fukuyama with the "end of history" understood as the triumph of Lockean liberalism.

The main value of returning to Hegel in our own era is not to see how we are progressing toward the end of history and the final flowering of freedom and reason, but to consider, on the contrary, how the twentieth century has blown apart the synthesis that Hegel believed was imminent after the Jacobin Terror of 1793 when the worst horrors of modernization were supposedly past. Looking back to that first revolution for transcending liberalism, we can only see modernity in the twentieth century as a series of sharp rifts and chasms, not as a lockstep progression of reason and freedom. All the contradictory forces that Hegel thought had been at least implicitly reconciled in 1806 blew apart in the twentieth century and persist or are even intensifying now: religious fanaticism, tribal rivalries and hatreds, uncontrolled technological might, fascism of the Left and Right, romantic narcissism versus arid proceduralism. Peace between the two modernist superpowers did not result in the dialectical supersession of the sources of modern alienation and hostility, but has been succeeded by the war against terrorism, genocide in the Balkans and Africa, and a host of burgeoning demographic and economic catastrophes in the developing world.[4] The end of the cold war has not made the world smaller and more homogeneous—the essence of Fukuyama's interpretation of Kojève's interpretation of Hegel—but larger, more fragmented, and arguably more dangerous. The time has come to try to think through how we have arrived at this dangerous place; to think through exactly what has been happening on our sometimes wonderful, sometimes frightening modern journey since the French Revolution. The collapse of one particular outcome of the revolution of Science and Love—Marxism-Leninism—may allow us to trace other paths on the journey with greater retrospective clarity. Indeed, the full stakes and complexity of modernity may only now be dawning as we pass out of the twentieth century.

I

Hegel's diagnosis of the modern age has as much to do with Solzhenitsyn's kind of spiritual critique of modernity as it does with vindicating the Enlightenment and Lockean liberal democracy. Hegel thought the new age was synthesizing both dimensions, the spiritual and the liberal-democratic—the spheres of Love and Understanding, or, as political theorists might put it currently, the spheres of community and rights. But can we believe in this synthesis today? Our experience so far in the twentieth century has been of the increased polarization of secular modernization, on the one hand, and of a religious or sentimental yearning for wholeness on the other. We want autonomy and community; individual rights and "roots"; endlessly productive technological economies and "the earth"; the freedom to define our lives as individuals and "the goddess." Hegel diagnosed this schism as the opposition between Understanding (by which he meant the analytical empiricism and contractual political right of the Enlightenment) and Love (the realm of immanent communal intersubjectivity).[5] He believed that the near future would harmonize these contradictory yearnings for individualism and reconciliation. What is truly relevant about Hegel today, I believe, is not the "end of history," but his brilliance in penetrating this basic—and continuing—tension within modernity.

The revolution of Understanding and Love will not only not disappear, but may well intensify. For Marxism-Leninism was only one historical consequence of Hegel's diagnosis of this characteristic modern dichotomy. Just because Marxism-Leninism has been discredited and, it would appear, removed from world history in no way means that the feeling of alienation from liberal modernity out of which Marxism-Leninism originally sprang will go away. Indeed, a new post-Hegelian, postmodernist paradigm is emerging for expressing a series of distinct but interlocking dissatisfactions with the still-dominant liberal paradigm. This new paradigm differs from past forms of radical opposition to liberalism because it lacks a focus and an agenda for revolutionary political action at the level of changing regimes. Instead, it will be more of a cultural revolution *within* the liberal-democratic world, slowly corroding its ethos from within. Now that the Soviet alternative to liberalism has vanished, we will return to the tension between Understanding and Love that Hegel originally diagnosed, not as a political assault on liberal democracy from without, but as a cultural revolution continuing to unfold from within.

This new paradigm can be evoked by a favorite nostrum of middle-class activism in North America, "think globally, act locally." This slogan captures the dawning perception that, as the nation-state and its politics fade away, we experience only what is closest to us (work, family, neighborhood, advocacy group) and what is farthest from us ("I care about this planet"). One can group under it a series of lively and spreading social movements. Each of them begins by identifying liberal modernity as the source of its alienation and the impediment to its freedom and fulfilment. Each of them posits a golden age of the past free of alienation and oppression, a golden age of no limiting conditions on spontaneous happiness and self-expression. Each of them believes that one must combat the global paradigm of liberalism with its technological and capitalistic adjuncts in order to allow their particular local community to return to the unconditioned bliss of the origins. And yet, by returning to its own particular version of the golden age, each of these movements more or less consciously believes that the shattering of the predominant liberal paradigm will allow these different local groups to inaugurate a planet-wide blossoming of greater freedom and happiness.

Here are some examples: 1) The fascination with the age of "the goddess," an age allegedly preceding the rise of male-dominated Olympian Greek culture when authority was matriarchal. 2) The belief of "men's rights" groups that there was also a prehistorical golden age when men were more in touch with nature and themselves, including the reenactment of allegedly genuine tribal and shamanistic rituals. 3) The popularization of "the age of mankind," a prehistoric era prior to the emergence of civil and commercial culture which is a historical and anthropological fact, but also serves as a normative standard for urging people to return to a condition of greater harmony with the earth. 4) Environmentalism itself, which often dovetails with No.3 to suggest returning to or at least imitating the tacit wisdom of our primordial ancestors' harmony with the environment. The atavistic project to recover this harmony points the way to the complete transformation of existing modernity. 5) The peace movement of the 1980s, according to which the entire course of Western civilization has been aimed at the pursuit of technological and nationalistic power, whose resulting nuclear terror may shock us into an advance into a peaceful postmodern future, which would at the same time be a return to premodern innocence. 6) The "black Athena" scholarship that locates the true origins of Western civilization with the peoples of Africa and Egypt, with the implication that Western civilization appropriated this heritage and perverted it to serve exploitative ends. 7) The emergence of an "aboriginal international" made up of premodern communities that regard themselves as autochthonous, each one possessing an irreducibly unique culture, yet linked with one another around the world to combat imperialistic nationalism and preserve the environment.

Despite the enormous diversity among and within these social movements, there is a common thread. They all maintain that human life was originally not characterized by alienation and oppression. The golden age is one of harmony with the environment, peace between the genders and among peoples, without bourgeois property relations or competition. In the more extreme ideological formulations, Western civilization is a compendium of oppressions—technological, racist, sexist. Using the golden age of the unconditioned as a guide, we can aim for a future in which we return to the past, throwing off the shackles of the present. As ideologues of the peace movement were fond of saying, we need to "reinvent politics," "reinvent the world." Consequently, even though global technology is usually perceived in these ideologies as the summation of Eurocentric, logocentric domination, these movements often envision using its power for their own projects of benign transformation. Technology may lead to disaster and oppression. But (as in Heidegger's late philosophy) it may also be turned against itself to release "the earth." Postmodernism is part of a cultural revolution for transforming liberal democracy from within, not a political revolution aimed at change at the regime level. The danger it presents is accordingly much more modest than that of Marxism-Leninism, but nonetheless quite real. This is the danger that, instead of focusing on concrete remedies to injustice (such as equal pay for equal work regardless of gender), a new generation of social and behavioral engineers will aim at the deconstruction and reconstruction of the human personality through psychotherapy and propaganda.

II

The postmodernist project of deconstructing and reconstructing the human soul is not confined to the Left. As another best-selling management guru has written, we must be ready to change "every nanosecond" for the sake of the dynamic fluidity required by global competitiveness. Just as the bourgeoisie unwittingly brings about proletarian consciousness when it pursues the maximization of profit to the exclusion of every other understanding of the human good and at the cost of corroding every substantive national and local community (and in these observations Marx was surely accurate), so our new version of "capital," the paradigm of global competitiveness, while preening itself on being the cutting edge of conservatism, unwittingly prepares the postmodernist nirvana when it seeks to subordinate and assimilate all other valid political and social concerns to its single imperative of dismantling the modern nation-state as an impediment to its revolutionary global mission.

The standoff at the 1997 Cairo conference on overpopulation may indicate how future struggles will unfold between what remains of liberalism and the Enlightenment, on the one hand, and religious and national tribalism on the other. On one side of the Cairo standoff was an emerging elite of international civil servants and social workers bent on curtailing the growth of the masses and inducing all peoples and cultures to accept liberalism's victory at the end of history—the policy heirs, so to speak, of Robert Owen, the Physiocrats, and the Philosophical Radicals, bent on reforming the masses for their own good. On the other hand, as Conor Cruise O'Brien has observed, we also saw what was perhaps the beginning of a revanchist alliance of Islamic fundamentalism and Christian conservatives (the heirs of the Counter-Reformation and nineteenth-century Romantic folk-nationalism).[6]

Apart from these competing visions in international relations, seemingly politically neutral advances in medical technology are also bringing about a postmodernist nirvana. Recent psychotropic drugs such as Prozac are not only recommended, as is entirely reasonable and desirable, for people suffering from clinical depression and other psychological disorders, but proselytized among the healthy as the way to create a new human being who is relentlessly upbeat, goal oriented, productive, well adjusted, and unerotic. In this vision of a medical utopia, one can do an end run around the virtues of character traditionally thought necessary to equip us to resist vice and to console us against failure and misfortune, because our chemistry can be fine-tuned to avoid the impulses that make these virtues necessary. A pill or syringe may deliver us to the golden age of the unconditioned more rapidly and more surely than earlier, cruder attempts to create utopias through revolutionary willpower such as Marxism. Why bother dismantling the positive, outward, and literal conditions of the political system when one can get to the heart of the matter and do what the Bolshevik and fascist regimes, despite ceaseless efforts at indoctrination and reeducation, never succeeded at doing: deconstructing and reconstructing the human soul? Such a chemically altered human being, if Prozac is anything to go by, will be the perfect embodiment of the postmodernist agenda—open, nonjudgmental, laid back, and non-hegemonic. But at the same time, and for the same reasons, such a person will be the perfect worker according to the global economic paradigm, easily adaptable to our ever more fluid, non-stratified "virtual" workplaces.

III

As I began by observing, ever since political philosophers began elaborating the concept of the common good, we have assumed that the civic association would be coextensive with a particular, autonomous polity, the nation-state being the locus for liberal democracy. But the slogan "think globally, act locally" is evocative of a profound change in the social, political, and economic reality of the late twentieth century that renders the very idea of the nation-state untenable. For today, capital is not merely multinational, but has no national basis at all. The archetypal American corporate executive of yesteryear, identifying what is good for America with what is good for his company, has been replaced by international money markets with no executive or even physical center. To paraphrase Foucault and Derrida, they are de-subjectivized networks of (financial) power, a free play of (financial) signifiers. The millions who contribute to them through pension funds, stocks, and bonds become the joint owners of thousands of enterprises from one hour to the next as their account managers search the world for a better point spread. Thus, as Robert Reich put it as the unprecedented global financial boom of the Clinton era got underway, the real question is not whether this global system is good for "us" in a given country. The real question now is, "who is 'us'?"[7]

Old-fashioned accounts of the bourgeois virtues such as that of Adam Smith assumed that a talent for commerce could be placed at the service of the common good of one's country, and that the virtues of diligence, sobriety, and probity required by commerce were themselves best instilled through the character formation that comes from belonging to a distinct civic association. No major philosophical exponent of liberal democracy and free enterprise ever advocated a life of unbridled moneymaking and materialism. On the contrary, it was always held that an education in moral character was needed if individual liberties were not to degenerate into vice. Smith is famous for formulating the argument that what had traditionally been regarded as private vice—the pursuit of profit through commerce—engenders public virtue. But Smith's endorsement of free enterprise economics presupposes educating the "inner man" in the moral and intellectual virtues that prevent us from being totally absorbed in moneymaking. According to Smith, people will not treat each other in a decent and law-abiding manner in their commercial relations unless those relations are guided by a wider moral training of our capacities for reason and sympathy.[8]

But economic globalization appears to be snapping the perhaps always fragile link between civic character and capitalism. To the extent that it forsakes the nation-state, global capitalism severs its link with even the rather qualified Lockean and Jeffersonian adaptations of classical virtue to modern individualism. That "worldly asceticism" which R. H. Tawney identified as the characterological core of bourgeois civilization—its virtues of thrift, honesty, diligence, steadiness, and probity—is considered to be as square and retrograde by contemporary management gurus as it was by Sixties hippies.[9] Global investment, technological R&D, the search for low-cost labor—the whole agenda of "competitiveness" that has summed up much of what is vital in parties that call themselves conservative today—are every bit as impatient of constraints by the old structures of the nation-state, and by the old structures of linear reasoning, as are deconstructionists or radical feminists. Capitalism has been transformed from a system of national elites of the managers of primary production into a global elite of information processors. Class divisions within nation-states are giving way to global class divisions between information processors, technicians, and laborers. This process unfolds in conjunction with a decentering of capital as it departs its traditional stewards in the nation-state and is dispersed into an endlessly fluid and mobile global environment. The same longing to burst the restraints of the old grammar and logic, the longing for the unconditioned, alike drives millenarian environmentalism, particle-laser weapons systems, and Disney World, where the goal is (as Umberto Eco has observed) to create a simulation of anything that has ever happened or ever could happen.[10] Laser technology, whether it serves Mickey Mouse or a missile defense system, is the ultimate realization of Derridean *"différance,"* a free play of signifiers in which no traditional ethical or logical restraint can be allowed to interfere with technology's infinite plasticity and power of creation.

What I term the longing for the unconditioned characterizes a host of movements dissatisfied with the liberal status quo. These movements are also attracted to post-Hegelian (which is to say Heideggerian) ontology—the longing for non-reifying discourse, a desubjectivized life world, and Derridean *"différance."*[11] This drive to go behind the copular "is," behind the constraints of linear logic and causality, is what happens when you attempt to remove Hegelian Understanding from Love—when you attempt to liberate the longing for wholeness from any reliance on an analytically and politically stable conception of permanent duties and rights. And yet precisely this same drive for deconstruction and intersubjectivity—the dream of living in a world without alienation, obligation, or constraint—lies behind the most advanced processes of contemporary technology and the capitalism it serves. What better example is there of this than the widespread addiction of the educated elites to the World Wide Web? Here is the perfect postmodernist community, actualized by the most advanced modern communications technology, a communications system originally developed by the Pentagon as a fail-safe network in the event of nuclear war. It perfectly crystallizes the contemporary cant of community, communities made up of people who in truth share little in common except for some single biological or ideological trait abstracted from the welter of obligations and duties that make up the warp and woof of real people's lives. One can "communicate" on the Web in complete invisibility and anonymity, a furtive, onanistic projection of an empty self upon other empty selves, dispensing with the inconvenience of other bodies and the souls that inhabit them, and so dispensing with the age-old need to talk to others, to try to love or at least understand them, which presupposes developing one's own virtues so as to make oneself lovable or at least intelligible.

The new world dreamt of by both postmodernism and global capitalism is a world without vices or virtues, a world where nothing need ever constrain us, even the limitations of syntax and predicative reasoning. Indeed, the coming golden age can only be evoked by its indifference to the laws of logic and rational discourse. The irony of the West at the beginning of the new millennium is that technological capitalism itself is creating the desubjectivized life world longed for by postmodernism. Whether it be through postmodernist architecture, chemical-based microprocessing, or the fantasies of cyberpunk, the straight line of Newtonian physics and its political correlation in the universal rights and institutions of the nation-state is everywhere giving way to the free happening of decentred Heideggerian Being. And as this global alliance of Left and Right unfolds, that great Victorian holdover and last haven of the old politics, the nation-state, appears increasingly unable to serve as a focus for retarding or limiting this process in the name of that autonomous rights-bearing subject that was the glory of the Enlightenment. This liberal subject—a blend of Puritan, Locke, Kant, and Hegel— sustained modernity for two hundred years with its independent-mindedness, godliness, and love of learning. But it now seems ever more peripheral to capitalism's most radical unfolding, as Bill Gates, today's Jay Gould in a pastel pullover, shows us "the way ahead." The question still emerging is how we can find new bearings for virtue and humanity as the world dissolves into interlocking processes of global technology and atavistic tribalism.

The moral and intellectual resources of the West are still strong and deep. Often they need only to be remembered. We need only try to articulate clearly for ourselves the way we already, for the most part, try to live. If we reject postmodernism's spurious invocation of premodern communitarianism and attempt to return intellectually to older teachings about politics and morality—always conscious of their limited applicability for the present—we find there much of the common sense that most people still live their lives by. All human beings (women, men, minorities) have the capacity to rise above their base impulses and cultivate the virtues of justice, generosity, friendship, gratitude, obligation, and citizenship. All human beings just as surely will at times give in to their vices and disgrace themselves or do harm to others—some only occasionally, others more frequently. The capacities for virtue and vice are distributed equally, on an individual basis, throughout the human species—both genders and all peoples. The art of politics is to encourage people to be good, while persuading—and, as a last resort, constraining—people to eschew vicious behavior. Human beings cannot be purged of their passions and prejudices, but those energies can be directed away from vice and toward virtue. Through education, we can try to sublimate aggressiveness and ambition into a sense of personal and civic honor that derives its self-esteem from being good and suffers pangs of shame and remorse over being bad.

Notes

1. Peter F. Drucker, "The Age of Social Transformation," *The Atlantic Monthly,* November 1994.

2. George Soros, *Soros on Soros* (London: John Wiley, 1995).

3. Francis Fukuyama, "The End of History," *The National Interest*, Summer 1989; Alexandre Kojève, *Introduction to the Reading of Hegel*, trans. James H. Nicholls Jr. (New York: Basic Books, 1968).

4. Consider, for example, Michael Ignatieff, *The Warrior's Honour* (London: Viking, 1998), 3–8, 53–61.

5. G. W. F. Hegel, *Phanomenologie des Geistes*, Hoffmeister ed. (Berlin: Meiner, 1952), 17–19; *Early Theological Writings*, trans. Knox (Philadelphia: University of Pennsylvania Press, 1965), 303–13.

6. Conor Cruise O'Brien, *On the Eve of the Millennium* (Toronto: Anansi, 1994), 11–18.

7. Robert Reich, *The Work of Nations* (New York: Vintage, 1992), 301–13.

8. Adam Smith, *The Theory of the Moral Sentiments*, in *Adam Smith's Moral and Political Philosophy*, ed. H. W. Scheider (New York: Hafner, 1948).

9. R. H. Tawney, *Religion and the Rise of Capitalism* (New York: Mentor, 1954).

10. Umberto Eco, *Travels in Hyper-Reality* (New York: Continuum, 1986).

11. Consider Jacques Derrida, *De la grammatologie* (Paris: Minuit, 1967), 16, 95; Michel Foucault, *Les Mots et les choses* (Paris: Gallimard, 1966), 342–45.

2

Communication versus Obligation

The Moral Status of Virtual Community

Darin Barney

> "We are assured that the world is becoming more and more united, is being formed into brotherly communion, by the shortening of distance, by the transmitting of thoughts through the air. Alas, do not believe in such a union of people."
>
> —from the homilies of the Elder Zosima, in Fyodor Dostoevsky's *The Brothers Karamazov* (1880).

Among the historical benefits of digital communications media is that they reveal and clarify the essence of technology. To believe, as we often do, that technologies are simply neutral instruments engaged in the production of material objects is to misunderstand a central condition of the modern European and American experience. Technologies are indeed productive, but along with objects they also produce certain

21

ways of being in the world and, conversely, the absence of certain other ways. That is to say, while technologies as instruments produce objects, technology as practice participates in the production of subjectivity. The word *production* (as opposed to *determination*) should be noted in the preceding sentence, as should the word *participates*. Technology, on its own, does not determine subjectivity wholly or outright, but every technology, in the context of an array of social, political, and economic conditions from which it cannot be separated, participates in producing human subjects in the world. That the essence of technology resides in its practical rather than its instrumental functions was decisively argued by Martin Heidegger at the apogee of technique in the middle of the twentieth century, and has been confirmed by most serious philosophers of technology writing since, including Heidegger's critics.[1] Heidegger's own way of expressing the essential character of technology was to say that technology, regardless of what it yields in its function as instrument, *enframes*.

Communications media reveal the essence of technology as enframing with great clarity, because their role in producing material objects is not always obvious. However, their role in producing and representing human relationships implicates them immediately in the constitutive practices of human subjects. This is especially the case with digital media of interpersonal communication, which, along with a vast array of objective, material effects, also quite clearly reconfigure, produce, and reproduce particular social, political, and economic practices to the relative exclusion of others. It is not always obvious what sorts of concrete objects digital instruments yield; it is readily evident that as technologies they produce "ways of being in the world," and subjects who are ready to be that way, and not ready to be other ways. It is in this light that I wish to consider the technological phenomenon of "virtual community." Digital communications media have many practical implications. In what follows I will argue that among that which is produced by this technology is a practice of community that is emptied of obligation and, so, drained of the moral attribute that distinguishes community from other types of relationships in civil society. In the particular social, political, and economic context in which they are situated (i.e., in liberal-democratic, high-technology capitalism), digital network technologies participate in producing virtual community—which is to say they help to produce community without moral obligation, and to reproduce the voluntarism essential to the contemporary liberal ethos.

Virtual Communities, Communication, and Interests

I should make clear from the outset that the subject of this investigation is the idea of *virtual community*, and not the practices of community or civic networks. So-called virtual communities exist entirely on-line: they are extrageographical, nonlocalized aggregations of individuals whose interaction is carried out exclusively across computer networks, via their participation in electronic mailing lists, multiple-user domains, chat and bulletin-board services and discussion groups. Here, the network *is* the supposed community and the community is a network. Community or civic networks, on the other hand, arise when network technology is used as an instrument of communication and information distribution by an already existing, geographically localized, off-line community, typically in an effort to enhance social participation and access to community goods or services. The distinction is crucial. The practices of civic networking assume that network technology can be *used by* communities; the idea of virtual community assumes that digital networks can *be* communities. Civic networking does not exhaust the manner in which digital technology confronts communities, and it is not certain these practices ensure a beneficial outcome for communities in this encounter. The scholarship investigating this question and these practices is growing steadily.[2] Whatever the case, this is not the issue being addressed here. My concern is with the idea of a digitally mediated virtual community.

References to the idea and existence of virtual community and its derivatives abound in popular literature celebrating the emancipating onset of the digital age.[3] Among the earliest, and most influential, attempts to articulate this idea and give an account of its manifestation in practice is Howard Rheingold's *The Virtual Community*, in which he recounts his experience as a pioneer of the legendary Whole Earth 'Lectronic Link (WELL).[4] The WELL is a computer-mediated network of multiple discussion groups based in southern California that links participants from across the globe. Participants in these groups debate, exchange ideas and information, commiserate, and engage in small talk across a broad range of subjects—tales of intense relationships, emotional bonding, and strong attachments on the WELL have reached mythological proportions.[5] Rheingold's initial volume has become a touchstone in debates about the promise and perils of virtual community building. Advocates of virtual communities—including Rheingold himself—consistently point out that flight to digitally mediated relationships such as those enabled by the WELL and similar services is fueled by the neglect and decay of off-line, *real*, public, community, and civic life.[6] Critics of virtual communities charge that the ready availability of privatized social interaction in cyberspace serves to intensify, rather than alleviate, the decline of community life in off-line places—indeed, it is argued that network communities have arisen as a fatal, final solution to a problem of civic decay that has been accelerated by the penetration of network technology more generally.[7]

In an early approximation, Rheingold defines virtual communities as: "cultural aggregations that emerge when enough people bump into each other often enough in cyberspace . . . [a] group of people who may or may not meet one another face to face, and who exchange words and ideas through the mediation of computer bulletin boards and networks."[8] Subsequently, he adds: "Virtual communities are social aggregations that emerge from the Net when enough people carry on those public discussions long enough, with sufficient human feeling, to form webs of personal relationships in cyberspace."[9] In her book on the relationships and identifications at play in network-mediated Multiple-User Domains (MUDs), Sherry Turkle refers to these associations as "virtual communities," which she defines as "a new kind of community . . . in which we participate with people from all over the world, people with whom we converse daily, people with whom we may have fairly intimate relationships but whom we may never physically meet."[10] Activist and writer John Coates defines "online community" as combining "a group of people having common interests," who jointly adhere to the same "Terms of Service for use of an online service."[11] The Canadian government, referring to the "growing reality" of "virtual communities" has determined that "geography will no longer be an obstacle for people with something in common getting together"—implying that a virtual community is a group of commonly interested people who get "together" in some manner other than physically, probably digitally.[12]

There are numerous ways to define virtual community.[13] My purpose here is not to review them comprehensively, but rather to isolate two elements that figure consistently and centrally in accounts of what constitutes virtual community, and to consider what these constitutive elements in fact define, or fail to define. The first element commonly presented as constitutive of virtual community is interpersonal communication; almost all accounts of this phenomenon are premised on the assumption that the act of communication is not just important to, but is in fact the essence of, community. In this view, the act of communication is an essential and primary facet of community between individuals. Most thoughtful accounts acknowledge that communication is not a sufficient condition of community, but even these maintain that communication is indispensable, and therefore central to community. For example, Fernback admits: "Not all virtual social gatherings are communities."[14] She lists "personal investment, intimacy and commitment" as ancillary attributes necessary to turn communication into community. Nothing in this acknowledgment detracts from the underlying conviction that communication is essential to community, that community cannot exist *without* communication, and that communication occupies a central, privileged, almost determining position in the range of practices that constitute community. Despite the aforementioned caveat, Fernback herself is careful to assert that "communication is the *core* of community."[15] Similarly, though he is careful to assert that the two should not be equated, Derek Foster asserts that "communications serves as the *basis* of community. . . . Community, then, is built by a sufficient flow of 'we-relevant' information."[16] Shawn Wilbur lists "the experience of sharing with unseen others a space of communication," as first among seven definitive attributes of virtual community.[17] Finally, the phrase "virtual community" has entered the inaugural Oxford dictionary of Canadian English as "a group of users who communicate regularly in cyberspace."[18]

The second element typically presented as constitutive in most accounts of virtual community is shared interest. If communication is the essential and constitutive practice of community in virtual communities, then shared interests are the privileged content of that practice. It is tempting to say that interests are the exclusive content of network-mediated communicative practices—as one of the aforementioned advocates of virtual community writes: "[C]ommon interests are the *only* real reason that people get together online to communicate."[19] This is an overstatement, but only a slight one. There are certainly accounts of people communicating on-line about things other than material self-interest.[20] Barry Wellman and Milena Gulia, for example, assert that "[e]motional support, companionship, information, making arrangements, and providing a sense of belonging are all non-material social resources that are often possible to provide from the comfort of one's computer," and provide a number of examples of such dynamics in operation.[21] However, as these authors recognize, evidence of these practices remains anecdotal and sparse, while it is well established and generally conceded that interests of one sort or another are the primary driver of network-mediated communication.[22] Furthermore, while things such as emotional support, companionship, and membership are certainly nonmaterial, it is not clear that they are disinterested. The interests of people commiserating over a common illness, or seeking respite from loneliness via the Internet may not be strictly material, but they are nevertheless interests that, if unsatisfied, would likely lead to disengagement by the interested party. To use terms that will become crucial below: people engage in virtual community because they wish to, not because they must—that is to say, because they are interested, materially or otherwise, in doing so.

Thus, while material interests may not monopolize network-mediated virtual community, shared interests of one sort or another do overwhelmingly characterize it. It is also the case that "relationships [that] develop on the basis of communicated shared interests" are not only privileged in many accounts of virtual community, but also held up as one of the phenomenon's most compelling attributes.[23] As Internet pioneer J. C. R. Licklider, writing prophetically in the 1960s, put it: "Life will be happier for the on-line individual because the people with whom one interacts most strongly will be selected more by commonality of interests and goals than by accidents of proximity."[24] Wellman and Gulia describe digitally mediated discussion groups—which are customarily presented as paradigmatic virtual communities—as "a technologically supported continuation of a long-term shift to communities organized by shared interests rather than by shared place or shared ancestry."[25] It should be noted that these are the statements of people who find no reason to despair of what they describe. It is therefore uncontroversial to conclude that while it may not exhaust the activity of virtual communitarians, communication of shared interest is certainly central to these relationships, and it substantively defines the character of these associations—associations presented as embodying the spirit of community.

To sum up, the defense of virtual community qua community rests on the assumption that the alchemy of communication and shared interests yields a type of human association that can legitimately be called a community. I am prepared to concede that virtual communities exhibit these two qualities in high relief. The question remains as to what is at stake in an account and practice of community constituted by these elements rather than others.

Community and Moral Obligation

My argument is that digital technology participates in producing communities (and a supporting discourse of "community") that are empty of moral obligation, arguably the essential core of this designation as it has been traditionally understood, and the attribute that substantively distinguishes community from other forms of civil association. Arguments about the centrality of common moral obligation to community, and about the threat technological mediation poses to the possibility of such communities, are not new. Nevertheless, the accelerating rise to prominence of discourses and practices of virtual community recommend revisiting and clarifying these arguments. In this section I will outline briefly an account of community as constituted by mutual moral obligation, and consider the relationship of such obligation to communication and interests, in order to define precisely what is absent in virtual communities.

To posit obligation as essential to community is far from controversial. As Neil Postman has pointed out, the etymology of the word itself suggests this meaning: "community" combines *cum* for "together with" and *munis* for "obligation."[26] In a widely cited contemporary definition, Thomas Bender stipulates: "A community involves a limited number of people in a somewhat restricted social space or network held together by shared understandings and a sense of obligation."[27] Conservative moral and political philosophy typically casts common obligation, especially that derived from an inherited history and tradition, as both a necessity and particular virtue of community.[28] Markate Daly describes "fairly wide agreement" among communitarian theorists on the conclusion that "friendship or a sense of obligation, rather than self-interest, holds the members [of communities] together."[29] Even liberals rooted in the thought of Thomas Hobbes and John Locke readily affirm that obligation—albeit prudential obligation embodied in contracts and derived from individuals' interest in securing benefit, or avoiding painful penalty—defines the essential relationship of a political community.[30]

Mutual obligation can be considered an essential element of community because it is the nature of an obligation to bind, to hold people together. The English word *obligation* is derived from the Latin *obligare,* the root of which *(ligare)* means "binding." It is linked to words such as legislation (in which binding obligations are expressed in law) and loyalty (which expresses faithful observance of an obligation). In this view, community is conceived as a sort of association to which one is bound rather than for which one volunteers: volition commits, but it is obligation that binds. To the extent that community can be meaningfully distinguished from other forms of civil association it is obligation, and the character of that obligation, that provides the substance of the distinction.[31] To say that obligation distinguishes community from other forms of civil association is to say that obligation is the particular excellence or virtue of community. This is not to say that individuals always experience obligation as unambiguously pleasing or interesting; it is to say that the presence and observance of these often displeasing and uninteresting obligations delineates community from other types of civil society relationships. The phrase "civil society" is important here: there have been and are other forms of relationship besides community—one thinks immediately of the family in the private sphere and the state in the political sphere—in which some form of mutual obligation at least potentially prevails as a binding force. What I am suggesting is that one way to specify community theoretically from other forms of non-private, nonpolitical, civil society associations is to identify common obligation as its essential binding force. Numerous commonalties (e.g., identity, locality, language, religion, etc.) make community relationships easier to establish and maintain, as do any number of salutary norms (e.g., fairness, reciprocity, tolerance, etc.). None of these commonalties or norms, however, define community as against other sorts of civil society associations. The argument here is not that common obligation is the only ingredient of a successful community, but rather simply that an association cannot really be considered a community without it.

Human beings find many reasons to associate, some of which—including shared interests, shared location, shared identity, and common experience—are often contingent attributes of communities, which complement and support the obligations that bind the community together. Absent such obligations, however, people associated on this basis (i.e., as gun owners, neighbors, women of color, survivors of abuse) remain associates, but it is not clear why we would describe their associations as communities rather than as various forms of interest, geographic, identity, or affinity groups. It is only of late that we reflexively designate as a community any aggregation of people linked by interest, proximity, identity, or experience.

In the account I am presenting here, community is a substantive designation reserved for civil society associations in which members—regardless of whatever else they may share—are bound by a mutually observed obligation to one another. To press this line of thinking further, I would like to suggest that it is *moral* obligation in particular, and even more specifically the mutually observed moral obligation to regard one's fellows despite one's interests, that characterizes community theoretically as a distinctive form of civil association.[32]

In the context of human relationships, to regard is to give heed to, to take into account, and to let one's course be affected by others. The English word *regard* comes from the French *regarder*, which translates as "to watch," but also means to look closely, think twice, and take care. *Regarder* derives from *garder*, which means to look after, to guard, to care for, to protect. When we regard others we not only take them into account and allow ourselves to be affected by them, we are also careful with them, we look after them, we protect them. To regard is to think twice before acting—our first thought is typically for our own interests, our second thought is at least potentially for the interests of others or for the common good. There are many accounts given in the Western philosophical tradition of the source and character of the moral obligation being sketched here as the mutual obligation to regard. For a variety of reasons that are beyond the scope of the present discussion, I prefer the account given by Simone Weil, who writes: "Obligations . . . all stem, without exception, from the vital needs of the human being," and that "[t]here exists an obligation towards every human being for the sole reason that he or she *is* a human being, without any other condition requiring to be fulfilled, and even without any recognition of such obligation on the part of the individual concerned."[33] My purpose here is not to establish the source of the moral obligation to regard one's fellows but rather simply to suggest that the word *community* designates a civil association within the boundaries of which this obligation is observed and enacted. I think this respects, rather than strains, a fairly common sense of the word *community* and its meaning, and accurately reflects what has traditionally been thought to distinguish community from other forms of association in civil society. Put bluntly, a community has customarily been thought of as a place where people look out for each other, regardless of whatever else might join or separate them.

The moral obligation to regard one's fellows, observance of which distinguishes community in the account presented above, bears an interesting relationship to communication and shared interest, the two defining attributes of virtual community. Political obligations, linked as they are to obedience (i.e., the obedience of a subject to a ruler), have an overtly communicative character. The word *obedience* derives from the Latin *edire/audire* for hearing: to obey is to do what you are *told* to do. Moral obligations, by contrast, are observed (often silently) and acted upon rather than communicated, felt as the quiet prick of conscience rather than uttered as consent or heard as command. Members of a community defined by a mutual moral obligation usually know what to do, or what not to do, without having to be told. Habituation to obligation, rather than its declaration, is the mark of community whose bond is moral. Rational, transparent communication, so important to democratic citizenship and legitimate political obligation, is not a requisite of membership in a community defined by mutual moral obligation.[34] Indeed, one of the consistent complaints about communities of mutual moral obligation is that they are often noncommunicative, opaque, irrational, and seemingly arbitrary. Similarly, anyone who makes a point of repeatedly declaring their moral obligations (rather than quietly observing and meeting them) is usually seeking to evade or be compensated for them. And unlike social contracts that establish liberal democratic political states, the moral obligations that bind traditional communities do not require consent or agreement; indeed, it is the mark of a moral obligation that it binds despite consent or agreement. Sometimes the moral obligation to regard our fellows invokes a duty to communicate (i.e., to tell the truth), but in these cases communication results from the obligation attached to community membership—it does not constitute the community or what binds it. Thus, the relationship between communication (the heart of virtual community) and the moral obligation to regard one's fellows (the heart of the alternative conception of community sketched here) is at best contingent.

The relationship between interests and obligations is more complex. The English word *interest* combines the Latin words *inter* for "between" or "among" and *esse* for "being"—an interest is something that exists between or among people. This suggests the dual, contradictory nature of interests: being between, they both join and separate. Interests join when two or more people have a similar interest, they separate or distinguish when people's interests differ or conflict. Interests can bring people together, but it should be noted that common or shared interests produce associations only of a particular type. All interests are a function of appetite, whose nature it is to be particular, dynamic, and demanding of satisfaction. Put simply, interest is self-centred, and wanes when appetite changes or has no prospect of being satisfied. Thus, associations based on common or shared interests persist only so long as the individual appetites animating them continue with a reasonable prospect of satisfaction. When appetites shift, or when one self-interest is eclipsed by another that is not complementary, an association built upon these appetites and interests will tend to dissolve. Unless there exists a compelling reason to maintain the association even when the interest for which it was established is no longer present, strong, or fulfilled, associations based solely upon shared (i.e., communicated) interests will tend to be unstable. Indeed, this was Thomas Hobbes's great insight into the nature of social relationships based on contracted exchanges of self-interest. People will remain sociable, and can live together peaceably, only so long as they have an interest in doing so. However, the same self-interest that leads individuals to seek peace in common will seduce them to seek advantage over their fellows at the first opportunity, despite their communicated promise of civility. As Hobbes famously wrote, in the face of self-interest "nothing is more easily broken than a man's word," and the force of words is "too weak to hold men to the performance of their Covenants." Thus, he concluded that "[c]ovenants, without the sword, are but words, and of no strength to secure a man at all."[35] The single interest capable of joining people in a stable political association—in Hobbes's terms, a "Commonwealth"—is the overwhelming fear of corporal or mortal punishment by a common power, a fear experienced equally by every individual. Absent the sword, human associations founded on contracted, communicated interests are inherently vulnerable and unstable.

Interests associate people when they are shared, but only temporarily, as long as these interests do not change or find themselves unsatisfied, in which case they will tend to divide people, unless an ultimate interest in survival recommends otherwise. This expresses the form of instrumental calculation that defines the character of political obligation under modern social contracts. C. B. Macpherson has suggested that such calculations of Hobbesian self-interest form prudential obligations, which cannot be distinguished from moral obligations.[36] It should be noted that Macpherson reaches this conclusion only after endorsing Hobbes's implicit rejection of the possibility of any nonprudential, disinterested, irrational source of moral obligation. Regarding Hobbes's account of instrumentally rational, strictly prudential obligation, Macpherson writes: "He thought it the best that men were capable of without fraudulently bringing in religious sanctions, and he thought it more moral that men should stand on their own reason than that they should evoke imagined and unknowable deities or essences. He thought that his rational, albeit self-interested, obligation was as moral an obligation as could be found."[37] Thus, prudential obligation can stand in for moral obligation only when the latter is dismissed as mere fantasy. If, on the other hand, one can conceive of non-interest-based grounds for obligation that are also not fraudulent or imaginary, then the distinction between moral and prudential obligation remains meaningful. In this case, a prudential obligation is one derived from, and pursuant to the satisfaction of calculated self-interest; a moral obligation is one by which we are bound despite interest and prudential calculation. Prudential obligations bind us voluntarily to that in which we have an interest (including other people); moral obligations bind us to that in which we have no interest (including other people) whether we volunteer or not. In essence, moral obligations often constrain precisely that which prudential calculation otherwise recommends, and it is in this constraint of interests that their binding action is made manifest.

In this section, I have argued that moral obligation is only contingently related to interpersonal communication, and that it can be distinguished from obligation derived from prudential calculation of self-interest. I have also presented a theoretical account of the nature of community based on the moral obligation of mutual regard. I think that observation of the mutual obligation to regard others substantively distinguishes genuine communities from other forms of civil associations, such as those based on shared interest, identity, or location. In my view a community is an association in civil society that binds individuals who meet its demand of mutual regard: in a community, individuals do not always get to do what they have an interest in doing, because sometimes their regard for others requires them to moderate their self-interest, and to do what they might not otherwise choose to do. However, it should be acknowledged that such views are idiosyncratic. Ours is a culture in which communication is valued over conscientious regard, in which moral obligation is considered an oppressive and reactionary phantom of a bygone era, and in which the pursuit of calculated self-interest is under-stood as the essence of freedom. The actual communities in which we live bear little or no resemblance to the account I have presented above as theoretically definitive of that designation.

The Essence of Virtual Community

What sort of community is a virtual community? In a fine article on the politics of the Internet, Bruce Bimber distinguishes between "thick" communities, which collectively pursue goals beyond the sum of their members' private interests, and "thin" communities, which are merely associations of individuals whose private interests are complementary. Bimber writes: "[O]ur understanding of the content of social interaction on the Net gives little reason to think that community will be significantly enhanced on a large-scale. Building community in a normatively rich sense is not the same as increasing the amount of social talk, and there is good reason to think the latter will be the norm on the Net."[38] He concludes that while network media are likely to facilitate the prolifera-tion and operation of thin communities of complementary private inter-ests, they are unlikely to contribute to the constitution and maintenance of thick communities which cohere around a collective good. In Bimber's estimation, digital technologies mediate an accelerated pluralism of inter-est groups, but not necessarily a substantive deepening of community. Even those who see greater potential in virtual communities generally concede that the relationships they contain are more or less defined by the communication of shared interests. Wellman and Gulia, for example, affirm that on-line relationships in virtual communities "develop on the basis of communicated shared interests."[39] Add to this the facility of digital networks to link similarly interested communicators otherwise separated by vast distances (or, in some cases, by arbitrary, visually cued prejudices) and the shape of virtual community begins to emerge.

However, it is possible that neither the mediation of shared interest nor communication across geographic and social barriers constitutes the particular virtue of virtual communities. Instead, I would argue, what distinguishes virtual communities is their status as associations in which the binding moral force of the mutual obligation to regard others is largely absent. That is to say, the distinctive excellence of virtual commu-nities is that they present a perfect technological solution to the problem of community in a liberal, market society. As William Galston has char-acterized it, this "central dilemma of our age" is one of somehow rec-onciling the overwhelming cultural value placed on individual autonomy and choice in liberal market societies with the abiding need human beings feel for association with their fellows.[40] The solution in liberal societies—at least in the Lockian variants that prevail in the contemporary West—has been to construct civil and political associations on the market model, wherein participation is voluntary and revocable, and the only constraints on the play of individual interest within the association are derived from the freely given and also revocable consent of the members.[41] This is supported by the grounding of political obligation in rights rather than in right; and social identification in shared but relativist values rather than in common faith with the good.[42] Thus, in a liberal market society that poses individual autonomy as the highest value, civil association is permissible so long as its terms express rather than constrain individual liberty understood as freedom of choice. Of course, under such conditions, the fundamental restrictions imposed by a moral obligation that operates and binds despite particular interest (the kind imagined by illiberal democrats such as Rousseau) are untenable.

Virtual communities meet the conditions of human association in a liberal market society better than most other forms of community, not only because of their bedrock foundation in communicated interest, but also because they are technically biased against the fundamental constraints of moral obligation that sometimes operate to bind individual choice making. Virtual communities are technically suited to meet the conditions of voluntarism—membership/obligation based on consent; low entry and exit costs; nonprejudicial relationships—which, in ensuring compatibility between social ties and autonomy, fulfill the test of legitimacy for liberal, market associations.[43] No one is obliged to be part of a virtual community in which they have no interest or for which they do not volunteer. As noted virtual communitarian Amy Bruckman writes: "In an ideal world, virtual communities would acquire new members entirely by self-selection: people would enter an electronic neighbourhood only if it focused on something they cared about."[44] There is no suggestion that virtual communities fail to live up to this ideal. Such voluntarism is of course in strong contrast to communities built upon moral obligation, which sometimes compel the duties of membership despite expressed interests, and in any case are not bound primarily by consent. Indeed, it is definitive of virtual communities that they lack, as a function of their technical organization, the binding force of what Stephen Doheny-Farina has characterized as "extraordinary communal constraint."[45]

A second mark of the voluntarism of virtual community is the ease with which prospective and dissatisfied members can enter and leave it. While it is true that certain virtual community formations maintain admission requirements and controls, these are not typically very restrictive, and usually involve some kind of basic qualification (which is typically an

automatic function of a person's interest in a particular virtual community in the first place—for example, I am interested in joining the virtual
community of political scientists on-line only because I am a political
scientist, which, not incidentally, qualifies me to join), an agreement to
adhere to certain protocols or rules of discourse, or the payment of token
fees. More important is the ease with which a member can sever himself
from a virtual community when it no longer holds his interest. Network
technology—as proponents of virtual community consistently affirm—
not only favors voluntary relationships based on the exchange of mutual
interests, but also makes it very easy to abdicate these relationships and
establish others as interests change. It is true that self-interested individuals become strongly attached to virtual communities that arouse their
interests.[46] It is also true that the very same appetites, arousal of which
prompts strong attachment, will cause individuals to abandon their attachments when they are no longer satisfying. As Wellman and Gulia
write, in support of virtual community: "Computer-mediated communication accelerates the ways in which people operate at the centers of
partial, personal communities, switching rapidly and frequently between
groups of ties. People have an enhanced ability to move between relationships."[47] Harmless defection from voluntary social relationships when
they no longer suit one or another party's interests is, then, precisely
what network mediation is configured to enable.[48] Again, such fluid,
contingent, and ephemeral attachments distinguish virtual community
from forms of community bound by mutual obligations that are not so
easily evaded, at least not without moral consequence. Thinking of such
a community, Doheny-Farina writes: "It is not something you can easily
join. You can't subscribe to a community as you subscribe to a discussion
group on the net."[49] Presumably, neither can one escape—or "unsubscribe"—
from communal obligation in good conscience, as one can from a virtual
community. However, ours is not an age of communities of moral obligation, and the "easy-come, easy-go" ethic of virtual community comports perfectly with the imperative of voluntarism that directs human
association in liberal market societies.

A third element of virtual community embodies the view of equality as neutrality, which is crucial to market liberal imperatives regarding
free and voluntary association. As an opaque medium that enables participants to obscure their identity and, in particular, any visible social cues
that might evoke stigmatization and arbitrary exclusion, computer networks provide associates with a socially neutral ground upon which they
might engage in nonprejudicial, nonhierarchical relationships based on
mutual interest and merit. Additionally, to the extent it is possible to
construct multiple identities and personae for use in a range of on-line

contexts and situations, virtual communities present an opportunity for an explosive liberation of "selves" that can access a variety of relationships previously denied to one's "self" due to inescapable physiological attributes and the prejudices attached to them.[50] One might expect this emancipation of alternative subjectivity to yield a genuine pluralism in virtual communities, especially relative to what is often characterized as the oppressive conformity demanded by communities bound by moral obligation. It is not clear that this is the case. Just as free markets promise consumer variety but tend toward homogeneity, virtual communities are often populated by members whose identities bear a striking uniformity and perfection. It is possible that what I have characterized above as moral communities are more open to difference than are virtual communities unleashed from the moral obligation to regard others.[51] As Doheny-Farina writes: "In physical communities we are forced to live with people who may differ from us in many ways. But virtual communities offer us the opportunity to construct utopian collectivities—communities of interest, education, tastes, beliefs and skills."[52] Though he prefers the language of market "incentives" to communal obligations, Galston is even more persuasive on this point: "When we find ourselves living cheek by jowl with neighbours with whom we differ but from whose propinquity we cannot easily escape, we have powerful incentives to develop modes of accommodation. On the other hand, the ready availability of exit tends to produce internally homogeneous groups that may not even talk with one another and that lack incentives to develop shared understandings across their differences."[53] Nevertheless, the shedding of prejudicial limitations on interest and appetite—including moral obligations—fits well with a notion of community as essentially voluntarist, and as enabling rather than constraining of individual liberty and choice. Plato's ring of Gyges conferred invisibility and, with that invisibility, liberty. It also, of course, constituted a license for immorality.[54] Community bound by mutual moral obligation cannot coexist with technologically enabled license. Recognition and enforcement of moral obligation, tied as they are to notions of accountability and responsibility, are impossible without reliable and stable identification. In Bimber's phrasing, "Trust, social capital, and the shared norms of thick community do not grow well in the soil of anonymity."[55] But freedom of choice—the essence of modern liberty—does grow well in this soil, which also happens to be the soil in which virtual community thrives.

In these three respects—consent; ease of entry and exit; and the minimization of prejudice via anonymity and identity multiplication—virtual communities fulfill the conditions of voluntarism that enable social connection without infringing upon individual autonomy and freedom of choice. They are thus a perfect form of association for modern, liberal, market societies, precisely because they combine fellowship with an absence of the moral obligation to regard others, the presence of which could not other than violate the existential demands of the current liberal dispensation. It is in this perfection that the particular virtue or excellence of virtual community is located.

Conclusion

In his critique of virtual community, Joseph Lockard has written that "[i]n the midst of desire we sometimes function under the conceit that if we name an object after our desire, the object is what we name it."[56] Are virtual communities really communities? Perhaps a better question is this: What sort of community is produced by digital network technology? I have argued that digital media participate in producing communities built on the communication of shared interest, communities constitutionally biased toward relieving their members of the mutual obligation to regard others, the moral quality that I think confers substance and meaning upon the designation *community*, and which ultimately distinguishes communities from other forms of association in civil society. Phrased differently, I find that the particular virtue of virtual community is that it lacks the virtue that marks the moral excellence of community, namely, the obligation of mutual regard. In this, virtual community resonates deeply with the voluntarist ethos of contemporary liberalism, and institutionalizes a form of association that gratifies the human appetite for fellowship without threatening the sanctity of individual autonomy realized through unfettered freedom of choice. It is often assumed that growing numbers of people seek out virtual community to compensate for the paucity of communal commitment in the off-line world. The argument of this chapter suggests an alternative explanation: that, despite their decay, communities off-line (in which conscience is still pricked when confronted by homelessness) continue to demand too much of individuals in the way of moral commitments that restrict the free play of choice, and that virtual communities offer people a means of reaping the benefits of communicative association without paying the constraining price of mutual obligation.

　　To be sure, some virtual communities do and will exhibit high levels of regard and obligation among members—sometimes even higher than is now typical in many physical neighborhoods. But such cases are still exceptional: they do not represent the technological bias of virtual community toward communicated interests; nor do they embody the particular excellence of virtual community as an essentially voluntarist form of association. On these terms, virtual communities resemble free markets more closely than they do moral communities bound by mutual obligation and regard. There is certainly support for this comparison in the rhetoric that presents virtual community as a potentially healthy form of human association. Wellman and Gulia, for example, affirm that "the very architecture of computer-networks promotes market-like situations," wherein people "shopping around for support" engage in relationships that are "intermittent, specialized and varying in strength." Here, community is reduced to a store from which one can "obtain a variety of resources," and membership to the practice of "maintain[ing] differentiated portfolios of ties."[57] The question is, what are the moral consequences of a technology that produces communities that cannot be distinguished from markets? That is to say, what is at stake when our fellows appear before us as resources whose relative value we can estimate, accumulate, and discard, rather than as priceless beings bearing an irreducible moral dignity that commands our regard, if not our interest; when the routine practices of community membership resemble shopping more than they do looking after your neighbors; and when consumer choice replaces moral obligation as the locus of our common humanity?

Notes

Thanks to Edward Andrew, Mary Stone, Don Desserud, and Tom Goud for insightful conversation and critical commentary on the theme and substance of this essay.

1. Martin Heidegger, *The Question Concerning Technology and Other Essays*, trans. William Lovitt (New York: Harper and Row, 1977).

2. See, for example: Leslie Regan Shade, "Roughing it in the Electronic Bush: Community Networking in Canada," *Canadian Journal of Communication* 24:2 (Spring 1999): 179–98; Peter Kollock and Marc A. Smith, "Communities in Cyberspace," in *Communities in Cyberspace*, ed. Marc A. Smith and Peter Kollock (New York: Routledge, 1999); Roza Tsagarousianou et al., *Cyberdemocracy: Technology, Cities, and Civic Networks* (New York: Routledge, 1998); Roger Gibbins and Carey Hill, "New Technologies and the Future of Civil Society," paper presented to the Annual Meeting of the Canadian Communication Association, Ottawa, Ontario, 31 May 1998; Philip Agre and Douglas Schuler, eds., *Reinventing Technology, Rediscovering Community: Critical Explorations of Computing as a Social Practice* (Greenwich Conn.: Ablex, 1997); Jay Weston, "Old Freedoms and New Technologies: The Evolution of Community Networking," *The Information Society* 13:2 (1997): 195–201; Douglas Schuler, *New Community Networks: Wired for Change* (Reading, Mass.: Addison-Wesley, 1996); Barry Wellman et al., "Computer Networks as Social Networks: Collaborative Work, Telework, and Virtual Community," *Annual Review of Sociology* 22

(1996): 213–38; Steve G. Jones, ed., *Cybersociety: Computer-mediated Communication and Community* (Thousand Oaks, Cal.: Sage, 1995); and Douglas Schuler, "Community Network: Building a New Participatory Medium," *Communications of the ACM* 37:1 (January 1994): 39–51.

3. For example: Peter Drucker, "The Age of Social Transformation," *Atlantic Monthly* 274:5 (November 1994): 53–80; Paul Hoffert, "The Bagel Effect," *Globe and Mail*, 14 January 1998, A11; Bill Gates, *The Road Ahead* (New York: Viking, 1995); Lawrence K. Grossman, *The Electronic Republic: Reshaping Democracy in the Information Age* (New York: Viking, 1995); William Mitchell, *City of Bits: Space, Time, and the Infobahn* (Cambridge: MIT Press, 1995); John Perry Barlow, "Is There a There in Cyberspace?" *Utne Reader*, March/April 1995, 53–56; George Gilder, *Life After Television: The Coming Transformation of Media and American Life* (New York: Norton, 1994).

4. Howard Rheingold, *The Virtual Community: Homesteading on the Electronic Frontier* (New York: Addison-Wesley, 1993).

5. See Katie Hafner, "The Epic Saga of the WELL," *Wired*, May 1997, 98–142.

6. Rheingold, *The Virtual Community*, 6. See also Barry Wellman and Milena Gulia, "Virtual Communities as Communities: Net Surfers Don't Ride Alone," in *Networks in the Global Village:Life in Contemporary Communities*, ed. Barry Wellman (Boulder: Westview, 1999), 332.

7. See, for example: Stephen Doheny-Farina, *The Wired Neighbourhood* (New Haven: Yale University Press, 1996). See also Joseph Lockard, "Progressive Politics, Electronic Individualism, and the Myth of Virtual Community," in *Internet Culture*, ed. David Porter (New York: Routledge, 1997), 219–31.

8. Howard Rheingold, "A Slice of Life in My Virtual Community," *The Electronic Frontier Foundation's Extended Guide to the Internet*, 1994, http://www.lpthe.jussieu.fr/DOC_HTML/eeg/eeg_260.html (September 1999).

9. Rheingold, *The Virtual Community*, 5.

10. Sherry Turkle, *Life on the Screen: Identity in the Age of the Internet* (New York: Simon and Schuster, 1995), 9–10.

11. John Coate, "Cyberspace Innkeeping: Building Online Community," January 1998 version, available online at http://www.sfgate.com/~tex/innkeeping (Sept. 1999). Originally published in Agre and Schuler, eds., *Reinventing Technology*.

12. Canada, *Building the Information Society: Moving Canada into the 21st Century* (Ottawa: Ministry of Supply & Services, 1996), 3.

13. For a good review of many of them, see: Steve Jones, "Information, Internet, and Community: Notes Towards an Understanding of Community in the Information Age," in *Cybersociety 2.0: Revisiting Computer-Mediated Communication and Community*, ed. Steve Jones (Newbury Park, Cal.: Sage, 1998). See also Jan Fernback, "There is a There There: Notes Toward a Definition of Cybercommunity," in *Doing Internet Research: Critical Issues and Methods for Examining the Net*, ed. Steve Jones (Thousand Oaks, Cal.: Sage, 1999), 203–20.

14. Fernback, "There is a There There: Notes Toward a Definition of Cybercommunity," 216.

15. Fernback, "There is a There There: Notes Toward a Definition of Cybercommunity," 213 (emphasis added).

16. Derek Foster, "Community and Identity in the Electronic Village," in *Internet Culture*, ed. David Porter (New York: Routledge, 1997), 24–25 (emphasis added).

17. Shawn P. Wilbur, "An Archaeology of Cyberspaces: Virtuality, Community, Identity," in *Internet Culture*, ed. David Porter (New York: Routledge, 1997), 13.

18. Katherine Barber, ed., *The Canadian Oxford Dictionary* (Toronto: Oxford University Press, 1998), 1622.

19. John Coate, "Cyberspace Innkeeping: Building Online Community," (emphasis added). See also Wellman et al., "Computer Networks as Social Networks: Collaborative Work, Telework, and Virtual Community," 224, 231.

20. See Rheingold's account of "The Heart of the WELL," *The Virtual Community*, 17–37.

21. Wellman and Gulia, "Virtual Communities as Communities," 338–39.

22. See Pippa Norris, "Who Surfs? New Technology, Old Voters & Virtual Democracy," in *Democracy.com? Governance in a Networked World*, ed. E. C. Kamarck and J. S. Nye Jr. (Hollis, N.H.: Hollis Publishing, 1999), 77–82.

23. Wellman and Gulia, "Virtual Communities as Communities," 352.

24. As quoted in Andrew L. Shapiro, "The Net that Binds: Using Cyberspace to Create Real Communities," *The Nation*, 21 June 1999. Available online at http://www.thenation.com/issue/990621/062shapiro.shtml (Sept. 1999).

25. Wellman and Gulia, "Virtual Communities as Communities," 336.

26. See Neil Postman, *Technopoly: The Surrender of Culture to Technology* (New York: Knopf, 1992).

27. Thomas Bender, *Community and Social Change in America* (Baltimore: Johns Hopkins University Press, 1982), 7–8.

28. See, for example: Alasdair McIntyre, *After Virtue* (Notre Dame: University of Notre Dame Press, 1981); Michael Oakeshott, *Rationalism in Politics and Other Essays* (Indianapolis: Liberty Press, 1991); and George Grant, "An Ethic for Community," in *Social Purpose for Canada*, ed. Michael Oliver (Toronto: University of Toronto Press, 1961).

29. Markate Daly, *Communitarianism: A New Public Ethics* (Belmont, Cal.: Wadsworth, 1994), xv.

30. For a discussion of the place of obligation in this tradition, see: C. B. Macpherson, *The Political Theory of Possessive Individualism* (Oxford: Oxford University Press, 1962), 70–87; and Carole Pateman, *The Problem of Political Obligation: A Critique of Liberal Theory* (Berkeley: University of California Press, 1979).

31. For a complete, and classic exploration of the distinction between communal and contractual social relationships, see Ferdinand Tonnies, *Community and Society: Gemeinschaft und Gesellschaft*, trans. C. P. Loomis (East Lansing: Michigan State University Press, 1964). For an excellent contemporary treatment of the relationship between mass communications and community

understood in these terms, see James R. Beniger, "Personalization of Mass Media and the Growth of Pseudo-Community," *Communication Research* 14:3 (June 1987): 352–71.

32. For a concise survey of the terrain of obligation in its many forms, see John Ladd, "Legal and Moral Obligation," in *Political and Legal Obligation*, ed. J. R. Pennock and J. W. Chapman (New York: Atherton, 1970). See also A. John Simmons, *Moral Principles and Political Obligations* (Princeton: Princeton University Press, 1979).

33. Simone Weil, *The Need for Roots: Prelude to a Declaration of Duties Toward Mankind* (Boston: Beacon, 1952), 7, 4–5.

34. On the communicative aspects of democratic rationality and legitimacy, see Jürgen Habermas, *Between Facts and Norms: Contributions to a Discourse Theory of Law and Democracy*, trans. William Rehg (Cambridge: MIT Press, 1996).

35. Thomas Hobbes, *Leviathan* (London: Penguin, 1985), 192, 200, 223.

36. See Macpherson, *The Political Theory of Possessive Individualism*, 72–74.

37. Macpherson, *The Political Theory of Possessive Individualism*, 73.

38. Bruce Bimber, "The Internet and Political Transformation: Populism, Community, and Accelerated Pluralism," *Polity* 31:1 (Fall 1998): 148.

39. Wellman and Gulia, "Virtual Communities as Communities: Net Surfers Don't Ride Alone," 352.

40. William Galston, "(How) Does the Internet Affect Community? Some Speculation in Search of Evidence," in *Democracy.com? Governance in a Networked World*, ed. E. C. Kamarck and J. S. Nye Jr. (Hollis, N.H.: Hollis Publishing, 1999), 45–61.

41. On the "self-assumed" character of obligation in the context of liberal individualism, see Pateman, *The Problem of Political Obligation*, 11–36.

42. On rights and values in contemporary liberal discourse, see Edward Andrew, *Shylock's Rights: A Grammar of Lockian Claims* (Toronto: University of Toronto Press, 1988) and *The Genealogy of Values: The Aesthetic Economy of Nietzsche and Proust* (Lanham, Md.: Rowman and Littlefield, 1995).

43. See Galston "(How) Does the Internet Affect Community?" 48.

44. Amy Bruckman, "Finding One's Own in Cyberspace," in *Composing Cyberspace: Identity, Community, and Knowledge in the Electronic Age*, ed. Richard Holeton (Boston: McGraw-Hill, 1998) 175.

45. Stephen Doheny-Farina, *The Wired Neighbourhood* (New Haven: Yale University Press, 1996), 5.

46. According to Wellman and Gulia: "[P]eople's allegiance to the Net's communities of interest may be more powerful than their allegiance to their neighbourhood communities because those involved in the same virtual community may share more interests than those who live on the same block." Wellman and Gulia, "Virtual Communities as Communities: Net Surfers Don't Ride Alone," 352.

47. Wellman and Gulia, "Virtual Communities as Communities: Net Surfers Don't Ride Alone," 356.

48. Bimber, "The Internet and Political Transformation," 151.

49. Doheny-Farina, *The Wired Neighbourhood,* 37.

50. See Sherry Turkle, *Life on the Screen.*

51. For a contrary account of the relationship between community and diversity or difference, see Iris Marion Young, "The Ideal of Community and the Politics of Difference," in *Feminism/Postmodernism,* ed. Linda Nicholson (New York: Routledge, 1990).

52. Doheny-Farina, *The Wired Neighbourhood,* 16.

53. Galston, "(How) Does the Internet Affect Community?" 52.

54. Plato, *Republic,* trans. G. M. A. Grube (Indianapolis: Hackett, 1974) 359c–360d.

55. Bimber, "The Internet and Political Transformation," 150–51.

56. Lockard, "Progressive Politics, Electronic Individualism, and the Myth of Virtual Community," 225.

57. Wellman and Gulia, "Virtual Communities as Communities: Net Surfers Don't Ride Alone," 353.

3

Technology and the Great Refusal

The Information Age and Critical Social Theory

Bernardo Alexander Attias

O n October 18, 2000, the richest man in the world addressed an audience of Silicon Valley's technological elite at the "Creating Digital Dividends" conference. The conference, according to its press release, was "aimed at finding private business solutions to the problem of the global digital divide.... The potential dividends include providing billions of people with access to educational and health services, boosting incomes and stimulating economic opportunity, and creating the tools to manage natural resources more sustainably."[1] According to one reporter, the conference "revolved around the thesis that technology can make both entrepreneurs and consumers out of people earning less than $1 a day."[2]

Bill Gates, "an ardent capitalist and technophile,"[3] surprised the audience by publicly questioning the free market dogma of the information age, which seems to hold that the unfettered expansion of the technology industries can meet all human needs, eventually putting food in the mouths of the world's poor. "Let's be serious," the multibillionaire implored. "Do people have a clear view of what it means to live on $1 a day? . . . There are things those people need at that level other than technology. . . . About 99 percent of the benefits of having (a PC) come when you've provided reasonable health and literacy to the person who's going to sit down and use it."[4] Gates sounded a little like Ralph Nader when he confronted the question of donating technology to the poor: "The mothers are going to walk right up to that computer and say, 'My children are dying, what can you do?' They're not going to sit there and like, browse eBay."[5] A *New York Times* reporter commented on the address that Gates has lost much of the faith that he once had that global capitalism would prove capable of solving the most immediate catastrophes facing the world's poorest people. He added that more philanthropy and more government aid—especially a greater contribution to foreign health programs by American taxpayers—are needed for that.[6]

Of course, neo-Marxist rhetoric coming from the world's wealthiest man (whose software company Microsoft paid no federal taxes in 1999)[7] is easily dismissed as a cynical public relations ploy.[8] But left-leaning rhetorics seem to be turning up in the strangest places. If Gates's comments seem out of place, it is because he explicitly confronts the myth of the revolutionary benefits of the information age, a myth that helped catapult Microsoft into the monopolistic position it currently occupies. This myth appears pervasive in contemporary popular advertising, which urges us to ride the wave of liberation promised by the information superhighway.

Every day in the United States we are bombarded by exuberant and optimistic representations of the vast promise of the "information revolution." Recent technological and economic developments, specifically the rise of computer technologies and the globalization of the world economy, are frequently hailed by politicians, pundits, and advertisers as signs of major paradigm shifts in global social and cultural relations.[9] We are frequently told that a truly global village is just around the corner; that the information age has brought us the end of destructive exploitation, sociopolitical conflict, and overall a more "connected" and hipper world.

Companies increasingly exhort us to buy their products using the language of revolution and thoroughgoing social change. Billboards shout that "The Revolution isn't Over: It has just begun!" atop the website address playboy.com. Images of 1960s rebellion are wedded to futuristic fantasies of technological togetherness in advertisements for cars, computers, and colas. William S. Burroughs floats around our television screens in Nike ads. Amidst glaringly Bolshevik imagery, one on-line computer store urges consumers to "Buy RAM Naked!"[10] Apple Computer, whose "1984" commercial featured rebel attacks on an Orwellian police state, now implores us to "think different" in a series of advertisements that defy both historical context and grammatical syntax. One Volkswagen ad sums up a current trend in advertising: "If you sold your soul in the 1980s, now's your chance to buy it back." Public Relations consultants Edward Grefe and Martin Linsky described the tactics of a "new corporate activism" in the PR industry: "to marry 1990s communication and information technology with 1960s grassroots organizing techniques."[11] Perhaps the ultimate expression of this ethos is the very existence of *Wired* magazine, which offers unadulterated technological boosterism masquerading as political transgression: "The net is merely a means to an end; the end is to reverse engineer government, to hack it down to its component parts and fix it."[12] Keith White elaborates:

> *Wired* has a simple message from which it never strays very far: computers are not implements of conformity, over-organization, and all those other evils of the 1950s; on the contrary, computers are fun. They are liberating. It will be a good thing—hell, let's go all the way: it will be a bona-fide utopia when we are all finally wired electronically together, the big culture conglomerates acting as intermediaries. Rebels with funky hairdos and rockin' attitudes will rule . . . .[13]

This rhetoric often seems infused with the spirit of social theorist Herbert Marcuse. Marcuse's notion of the "Great Refusal," a vaguely defined yet powerfully motivational image of revolutionary change that appears in both *Eros and Civilization* and *One-Dimensional Man,* has inspired various factions of the American left since the late 1960s. The image of the dispossessed "just saying no" to the society that exploits them has been embraced for a variety of purposes and causes, usually without much analysis of Marcuse's actual inquiries into the ills of advanced industrial society.

Marcuse's description of one-dimensional society decries the eminently reasonable and utilitarian form taken by totalitarian power in the twentieth century. The "comfortable, smooth, reasonable, democratic unfreedom"[14] with which Marcuse began his polemic on modern society is anything but the "spectre" haunting Europe with which Marx began his. The "great refusal" on which he ends the work is hardly the unambiguous endorsement of revolt that many on the Left have assumed, and it certainly contained none of the unbounded technological optimism so present in the discourse of the information revolution. Marcuse's refusal is as profoundly ambivalent as is his critique of industrial society; he warns us in the preface to his book of the "fundamental ambiguity" of the overall thesis of his critique of industrial society: on the one hand, he posits that advanced industrial society has the resilience and stability to absorb any threat of significant social change; on the other, however, the "forces and tendencies which may break this containment and explode the society" are always present alongside this resilience. As Harrell notes, Marcuse "is apparently only slightly consoled by the prospect of the great refusal; and that is an apparently judicial attitude if seen from the perspective of the mid-70s, since the abstract nature of the great refusal of the late 60s seems to have doomed it to failure, frustration, and 'consummation, in the mad antics of the Weathermen, dope fiends, and random violence.' "[15]

If the Great Refusal was poorly articulated in Marcuse's work, the contradictions of technological society that made it imperative were not. Marcuse clarified the problem of "one-dimensional thought" as a discursive co-optation of any threat to established conceptions of reality. Any ideas that "transcend the established universe of discourse and action are either repelled or reduced to terms of this universe."[16] For Marcuse, technological society transformed the class structure so that the traditional torch bearers of revolutionary change, the working class, are increasingly complacent thanks to bourgeois society's uncanny ability to satisfy the false needs it creates just enough to make people keep coming back for more.

It is instructive to note that at the same time the technophiliac obsession with the information revolution has taken a decisively Marcusian rhetorical turn, contemporary public and academic discourse has been proclaiming the death of Marx's critique of political economy. Since at least 1970, "futurists" have been declaring the end of the industrial age and the transition to an information-based economy.[17]

It is important to consider both the political implications of the rhetoric of the "information age" and the economics of domination that underlie the political structure of this so-called revolution. One might say that this rhetoric went into hyperdrive in the early 1980s with the publication of John Naisbitt's *Megatrends*.[18] It was certainly about that time that this rhetoric found a home in the American vocabulary as an unquestioned assumption as well as became a dominant motif of politicians and corporate leaders. Naisbitt had argued that "we now mass-produce information the way we used to mass-produce cars. In the information society, we have systematized the production of knowledge and amplified our brain-power. To use an industrial metaphor, we now mass-produce knowledge and this knowledge is the driving force of our economy."[19] The U.S.

Office of Technology Assessment picked up on this rhetoric in 1982, announcing as a *fait accompli* that "the United States has become an information society dependent on the creative use and communication of information for its economic and social well-being."[20]

Naisbitt mobilized this rhetoric against Marxism, calling for "a knowledge theory of value to replace Marx's obsolete labor theory of value."[21] Theodore Roszak points out, "It is difficult to see that writing like this (and the passage is typical) has any meaning at all, so deep are the confusions that underlie it."[22] Yet, whatever it does mean, it is clear that this kind of thinking has come to dominate both the corporate and the governmental landscape, at least when dealing with economic issues. Former secretary of labor Robert Reich, for example, gave an optimistic assessment of the American economy when forced to respond to media "leaks" about the Federal Reserve Board's policy of encouraging unemployment in order to keep business confidence high:[23]

> Yet the long-term data also affirm the potential for building a new middle class. Some of the fastest job growth in America is occurring among technicians, who defy the categories of the old economy. In my travels I've met vending-machine repair workers who use hand-held computers to identify problems and communicate with the home office, and managers in retail stores who tap personal computers to monitor sales and inventories. Some workers drive trucks equipped with computers and modems and make just-in-time deliveries. Others orchestrate sophisticated spreadsheet and graphics programs to create value in ways unimaginable a decade ago.[24]

Underlying this and similar statements is the assumption, most forcefully embodied in Naisbitt's statement above, that with the rise of the information age has come the end of production as a meaningful category of political economy.[25] Labor, it is contended, has undergone a radical transformation due to the rise of the technology industries and the increasing dependence of the international economy on "information." In Naisbitt labor undergoes a simple erasure; in Reich it is transformed somewhat more subtly but with consequences no less dramatic. Through what Doug Henwood has called a "reckless anecdotalism [that] would do Reagan or Bush proud,"[26] Reich argues for the rise of a "new middle class" of low-skilled laborers with gadgets.

Of course, the premise that the information age is dominated by the labor of "symbolic analysts" in an interdependent world economy has more than a grain of truth to it.[27] It's a grain that Marcuse noticed thirty-five years prior, when he discussed the "decisive transformation"[28] of the working class. Essentially, Marcuse argued that technological society had created layers within the lower classes due to fundamental changes in the quality of the labor experience. Where Marx had approached the labor of the working class as primarily physical labor, the new economy demanded and rewarded "skills of the head rather than of the hand; of the logician rather than the craftsman; of nerve rather than muscle; of the pilot rather than the manual worker; of the maintenance man rather than the operator."[29]

Marcuse quickly and correctly put his finger on the fundamental flaw in the optimistic discourse of mechanization of his day, which roughly mirrors contemporary information revolution discourse. The often repetitive tasks of the new knowledge worker are not much of a step away from the intensity of physical labor. Similarly, Reich's "symbolic analysts" are in no essential way better off than traditional laborers. As Henwood points out, "If their hand-held computers are designed correctly, they require no great skills—all the skill is programmed into the machine— and promise their operators no high wages."[30] Additionally, according to Marcuse, the dreary monotony of much symbolic labor has an additional benefit for the stability of advanced industrial society—it functions as a drug that lulls workers into enslavement to its machinic apparatus:

> The machine process in the technological universe breaks the innermost privacy of freedom and joins sexuality and labor in one unconscious, rhythmic automatism—a process which parallels the assimilation of jobs.[31]

It is worth briefly examining Marcuse's thesis in light of what Marx wrote in the notorious "Fragment on the Machine." This obscure passage from the *Grundrisse* on the effects of automation has been reexamined in light of its insights for an understanding of the "information age." Robert Reich's category of "symbolic analysts" is seen by some at least as corresponding to (if wildly distorting) the notion of "general intellectuality" which Marx began to develop in 1858.[32] While a thorough discussion of the "Fragment" is beyond the scope of this study, I would like to draw out a couple of Marx's insights on technology relevant to the current project. Marx wrote:

> Nature builds no machines, no locomotives, railways, electric telegraphs, self-acting mules, etc. These are products of human industry; natural material transformed into organs of the

human will over nature, or of human participation in nature. They are organs of the human brain, created by the human hand; the power of knowledge, objectified. The development of fixed capital indicates to what degree general social knowledge has become a direct force of production, and to what degree, hence, the conditions of the process of social life itself have come under control of the general intellect and been transformed in accordance with it. To what degree the powers of social production have been produced, not only in the form of knowledge, but also as immediate organs of social practice, of the real life process.[33]

It should be clear from this passage that Naisbitt's and Reich's easy movement from a labor theory of value to an information theory of value neglects any useful understanding of "information" as a social product that arises from real human relations rather than as a mysterious but quantifiable entity that is always conceived as ahistorical and asocial. Technologies are products of human industry, and the "information" they produce arises out of a set of social relations. The conditions and modes of production remain crucial in an analysis of the "information society."

One point may be obvious, but nonetheless must be stated: Reich's and Naisbitt's premise of an erasure of the significance of physical labor (and with it a Marxian understanding of the mode of production) is simply and utterly false. First of all, there are not two economies, one industrial and one "informational." The information economy is an industrial economy, at least so long as information machines need to be manufactured. As Roszak and others have demonstrated forcefully, manufacturing and labor have not been eliminated by the high technology economy.[34] Roszak writes:

In large part, the advent of the information economy means that our major corporations are rapidly retiring two generations of old capital or moving it abroad. As they do so, with the rich support of military contracts, they are liberating themselves from the nation's most highly unionized labor so that investment may be transferred into more profitable fields. High tech is not only glamorous; it pays off handsomely, especially if those who are collecting the profits are excused from paying the social costs that result from running down old industrial centers and disemploying their work force in favor of relocating in the mainly "right to work" (nonunion) Sunbelt states.[35]

So, while labor has not been eliminated by the rules of the new information economy, its international structure has been transformed. In particular, the division between the "haves" and the "have-nots" becomes more extensive and insurmountable. The class structure of Silicon Valley remains instructive as a microcosm of what is developing in the international arena. While software developers live in huge mansions and pull in six- to seven-figure annual incomes, fully one-third of the workforce in the valley earns less than $15,000 a year.[36] For most workers, Silicon Valley, according to Everett Rogers and Judith Larsen, "means low-wage, dead-end jobs, unskilled, tedious work, and exposure to some of the most dangerous occupational hazards in all of American industry. It is a dark side to the sparkling laboratories that neither barbecues, balloons, nor paid sabbaticals can hide."[37]

Yet this dark side has in fact been hidden in the increasingly optimistic rhetoric of the information age, although the veil of deception is often thin indeed. In 1996, Jerry J. Jasinowski, president of the National Association of Manufacturers, dismissed claims about corporate downsizing and the fate of the American worker with the following outright lie:

> Indeed, a wage boom has been underway for some time in many high-tech firms. Assembly-line positions in the technology sector now typically pay anywhere from $50,000 to $75,000 annually, including bonuses. And in part because of automation that has raised the skill-level required to perform all kinds of jobs on the factory floor, manufacturing workers in any field now earn an average of $40,000 annually. For companies like Cypress Semiconductor in San Jose, Calif., compensation is even higher: The average worker in this 1,900-person company, including line workers and receptionists, earns $93,000 a year including benefits.[38]

For the record: the average receptionist at Cypress Semiconductor in 1996 was subcontracted from a temporary agency at less than $25,000 a year, without benefits.[39] That statements such as Jasinowski's can be trotted out in the mainstream press without comment or challenge should give communication scholars pause when we hear about the "death of Marx" in the information age. For while American workers are fed lines like these from C.E.O.s,[40] the high technology industry is increasingly reliant on subcontracted, nonunionized, and illegal labor in this country as well as exploited "cheap" labor in large sectors of the "third world."[41]

In any case, both Marcuse's thesis of one-dimensional society and Marx's "Fragment on the Machine" point to a more profound critique of this discourse. It was Marx's contention that the machine, as a commodity, concealed the social power of labor with which it was invested. Workers who worked with machines, the "symbolic analysts" of Marx's day, effectively became the apparatuses of the machines, channels through which machines performed their subjectivity. There is a subtle twist to the process of commodification here in that subject and object change places; the worker becomes the channel through which the machine, or a machinic consciousness, exerts agency. Marx continued:

> The worker's activity, reduced to a mere abstraction of activity, is determined and regulated on all sides by the movement of the machinery, and not the opposite. The science which compels the inanimate limbs of the machinery, by their construction, to act purposefully, as an automation, does not exist in the worker's consciousness, but rather acts upon him through the machine as an alien power, as the power of the machine itself.[42]

The worker, in other words, becomes a part of the machine itself; s/he is reduced to the status of a cog in the teleological movement of the machinic apparatus. The social existence of human labor qua labor in this teleology is literally written out of the relationship. The human worker becomes the instrument of the "ghost in the machine"[43]—agency, creativity, and intention pass from human being to machine. Interestingly, even the ruling class is inexorably assimilated by the totalitarian logic of the machine. "Domination," Marcuse warned, "is transfigured into administration."[44] In the words of one clever Hollywood advertiser, "humanity is roadkill on the information superhighway."[45]

And yet, this becoming-machine of humanity is itself an entirely human process. Marx reminds us in the passage cited earlier that machines are "organs of the human brain, created by the human hand." Technology is not an autonomous force that acts on human beings; more accurately, human beings act on each other through technology. Human beings do not simply cede their subjectivity to autonomous machines; more profoundly, human beings actively participate in the performance of a seemingly inhuman subjectivity.

The phrase "information age" itself exemplifies this process at work. The word *information* has as its root the term *inform*, from the Latin verb *informare*, which means "to put into shape" or "to mold," and was often used to mean "to educate"—to mold other human minds as a way of bringing them to a higher social level. Yet its contemporary meaning seems to erase this actively inculcative property: "information" tends to be understood as an ahistorical, autonomous collection of data that exists independently of human activity. The notion of an era called "the information age" implies some prior event of momentous importance in the history of human progress that cannot be altered. Roszak compares the rhetoric of the "information age" to previous "ages" such as the "age of reason" and the "age of discovery." This kind of rhetoric functions to close off discussion on the meaning of the present by constituting a category as an inevitable historical process that must be adapted to rather than examined or challenged. "Unlike 'faith' or 'reason' or 'discovery,'" he writes, "information is touched with a comfortably secure, noncommittal connotation. There is neither drama nor high purpose to it. It is bland to the core and, for that very reason, nicely invulnerable. Information smacks of safe neutrality; it is the simple, helpful heaping up of unassailable facts. In that innocent guise, it is the perfect starting point for a technocratic political agenda that wants as little exposure for its objectives as possible."[46]

Ultimately, Marcuse found the neutral rationality of advanced capitalism's machinic apparatus its most troubling aspect. In the face of a system that so successfully interpolates its subjects and rationally explains away any threats to it, the only locus for the Great Refusal are "those whose life is the hell of the Affluent Society"—a lumpenproletariat that is "kept in line by a brutality which revived medieval and early modern practices."[47] This class forms a "substratum of outcasts and outsiders, the exploited and persecuted of other races and other colors, the unemployed and the unemployable. . . . Their opposition hits the system from without, and is therefore not deflected by the system; it is an elementary force which violates the rules of the game and, in doing so, reveals it as a rigged game. . . . The fact that they start refusing to play the game may be the beginning of the end of a period."[48]

Of course, Marcuse goes on to admit that "[n]othing indicates that it will be a good end." Critical theory is negative only because it wishes to "remain loyal" to those "without hope"—it ultimately "hold[s] no promise and show[s] no success." It is my suspicion that the Left's embrace of Marcuse's thesis of one-dimensional society and the valorization of the "Great Refusal" in particular seems ill-informed. One of the biggest problems is Marcuse's conception of power and resistance, which Foucault explicitly critiqued in *The History of Sexuality*. For Foucault, power and resistance do not operate externally to one another. Marcuse fundamentally seems to agree with Foucault's thesis that power circulates rather than being fixed—the movement from domination to administration is especially consistent with Foucaultian theories of power. Yet as Foucault points out, resistance operates from within the same context as power, from within the same machinic apparatus. Thus, resistance to power takes the same form as the exercise of power—multiple and heterogeneous rather than monolithic. "Hence there is no single locus of great Refusal," Foucault writes, "no soul of revolt, or pure law of the revolutionary. Instead there is a plurality of resistances."[49] In any case, what has become clear in the information age is that the rhetoric of the Great Refusal can be used to sell computers and fax machines as well as it can be used to articulate a critique of consumer capitalism. The irony, as Foucault wrote in a much different context, "is in having us believe that our 'liberation' is in the balance."[50]

Notes

1. Adlai Amor, "Gates, Other Industry Leaders to Address Digital Dividends Conference," Press Release from World Resources Institute (October 2000). Available: WWW URL: http://digitaldividend.org/press_exp_temp.asp?name=34.

2. Dan Richman, "Gates Rejects Idea of E-Utopia," *Seattle Post-Intelligencer*, 19 October 2000, A1.

3. Ibid., A1.

4. Ibid., A1.

5. Bill Gates, remarks to "Creating Digital Dividends" Conference (18 October 2000), quoted in Robert Scheer, "Gates Sends a Message: A Wired World Can't End Poverty," *Los Angeles Times*, 7 November 2000, B9.

6. Sam Howe Verhovek, quoted in Scheer.

7. Taking advantage of a tax loophole that allows companies to avoid income taxes on stock options they pay their employees, Microsoft and Cisco, among other large companies, didn't pay any taxes in 1999. See Kathleen Pender, "Giant Cisco Didn't Pay Any Federal Income Tax Businesses Get Break on Employee Stock Options," *San Francisco Chronicle*, 9 October 2000. Available: WWW URL: http://sfgate.com/cgi-bin/article.cgi?file=/chronicle/archive/2000/10/09/MN3707.DTL.

8. Gates does in fact use at least part of his time at the podium to extol the philanthropic spirit of the Microsoft corporation. "He said all the money recovered from pirated software in foreign countries—more than $50 million over the past five years—is donated to job training for the jobless and teachers" (Richman). The rhetorical displacement at work here is amazing—money extorted from businesses and organizations in third world countries under the guise of copyright enforcement by the notoriously litigious Business Software Alliance and then funneled into tax-free charities at home is successfully portrayed as an act of international philanthropy.

9. To be fair, there have been notable signs of meaningful paradigm shifts emanating from the technology industry. An important one is the remarkably well-informed advocacy of a more cooperative spirit of international capitalism embodied by real industry practices. Leading this charge are the users and advocates

of open source software and hardware. See, for example, Eric S. Raymond, *The Cathedral & the Bazaar: Musings on Linux and Open Source by an Accidental Revolutionary* (Cambridge, Mass.: O'Reilly & Associates, Inc., 1999); Chris DiBona, Sam Ockman, and Mark Stone, *Open Sources: Voices from the Open Source Revolution* (Cambridge, Mass.: O'Reilly & Associates, Inc., 1999); Richard M. Stallman, "Philosophy of the GNU Project" (Boston: Free Software Foundation 2000), available: WWW URL: http://www.gnu.org/philosophy/philosophy.html. While one can plausibly argue that the cooperative spirit advocated by the open source and free software movements constitutes a genuine shift in the landscape of international capitalist expansion, it is instructive to note that the emerging debate on these issues has so far taken place without questioning the assumptions of the dominant capitalist ideology. Nevertheless, open source advocates take the advertising trend discussed below to extremes. As one major open source advocacy site proclaims of the open source Linux operating system: "Linux—A Means to World Liberation!" (http://linux.com/).

10. MacGurus; on-line at http://macgurus.com.

11. Quoted in Sheldon Rampton and John Stauber, "Keeping America Safe from Democracy," *PR Watch* 5:3 (1998): 2.

12. Cited in Keith White, "The Killer App: *Wired* Magazine, Voice of the Corporate Revolution," in *Commodify Your Dissent: The Business of Culture in the New Gilded Age*, ed. Thomas Frank and Matt Weiland (New York: W.W. Norton, 1997), 46–56, 53.

13. Ibid., 54. It is interesting that in *Wired*'s two failed attempts to launch an IPO it described itself as a business offering "compelling, branded content with attitude" and bragged that "none of the Company's employees is represented by a labor union" (55). "Let's see," continues White. "Branded content with attitude, dream jobs, and a union-free workplace. You won't get paid much, and of course your benefits will probably suck, but everyone will respect your (branded) 'tude" (56).

14. Herbert Marcuse, *One-Dimensional Man: Studies in the Ideology of Advanced Industrial Society* (Boston: Beacon, 1964), 1. Hereafter ODM.

15. Bill J. Harrell, *Marx and Critical Thought* (Buffalo: State University of New York Press, 1976). Available: WWW URL http://www.sunyit.edu/~harrell/billyjack/marx_crt_theory01.htm. Harrell's quote, "consummation, in the mad antics of the Weathermen, dope fiends, and random violence," cannot be clearly attributed to Marcuse.

16. ODM, 12.

17. See, for example, Alvin Toffler, *Future Shock* (New York: Random House, 1970); Daniel Bell, "Notes on the Post-Industrial Society I," *The Public Interest* (Winter 1967): 24–35; Daniel Bell, "Notes on the Post-Industrial Society II," *The Public Interest* (Spring 1967): 102–18; Daniel Bell, *The Coming of Post-Industrial Society* (New York: Basic Books, 1973), 33–40; Daniel Bell, "The Third Technological Revolution and its Possible Socioeconomic Consequences," *Dissent* (Spring 1989): 164–76; and Zbigniew Brzezinski, *Between Two Ages: America's Role in the Technetronic Era* (New York: Viking Press, 1970). The

term *postindustrial*, popularized by Bell, actually dates back to the 1920s; see Arthur Joseph Penty, *Old Worlds for New: A Study of the Post-Industrial State* (New York: Macmillan, 1918), and Arthur Joseph Penty, *Post-Industrialism* (London: Arthur, 1922). On the genealogy of this term, see Armand Mattelart, *Mapping World Communication: War, Progress, Culture*, trans. Susan Emanuel and James A. Cohen (Minneapolis: University of Minnesota Press, 1994), 129.

   18. John Naisbitt, *Megatrends: Ten New Directions Transforming Our Lives* (New York: Warner, 1982). See also John Naisbitt and Patricia Aburdene, *Re-Inventing the Corporation: Transforming Your Job and Your Company for the New Information Society* (New York: Warner, 1985); and John Naisbitt and Patricia Aburdene, *Megatrends 2000: Ten New Directions for the 1990's* (New York: Avon, 1990).

   19. John Naisbitt, *Megatrends*, cited in Theodore Roszak, *The Cult of Information: A Neo-Luddite Treatise on High-Tech, Artificial Intelligence, and the True Art of Thinking* (Berkeley: University of California Press, 1994) 22.

   20. Office of Technology Assessment, *Information Technology and its Impact on American Education* (Washington, D.C.: U.S. Government Printing Office, 1982), cited in Roszak, 23–24.

   21. Quoted in Roszak, 22–23.

   22. Roszak, 23,

   23. Frederick Thayer pointed out the fundamental problem with Reich's line of reasoning in a subsequent letter to the editor: "Secretary of Labor Robert B. Reich's boastful summary of the Clinton Administration's 'work-force agenda' in 'The Fracturing of the Middle Class' (Op-Ed, Aug. 31) is political disingenuousness. It is not just 'Wall Street' that 'frets' that 'unemployment is too low to contain inflation.' Mr. Reich knows that the Administration, Congress and the Federal Reserve are determined to keep people out of work to depress wages and salaries. The creation of 2.5 million jobs in the last year is not 'progress,' and Mr. Reich knows it. He has repeatedly acknowledged that most of the new jobs are part time, low paid or both. The accepted policy to maintain at least 6 percent unemployment while labeling it 'full employment' insures that the number of people out of work increases as the work force grows. Mr. Reich knows this, and knows that it is a formula for social disaster. When will he propose a real jobs policy?" Frederick C. Thayer, "Don't Call Unemployment a Jobs Plan," *New York Times*, 7 September 1994, A22.

   24. Robert Reich, "The Fracturing of the Middle Class," *New York Times*, 31 August 1994, A19.

   25. That this assumption remains at the heart of U.S. labor policy is evident in current U.S. Labor Secretary Elaine Lan Chao's call for a new office of the "Twenty First Century Workforce." See Elaine L. Chow, "Remarks for Welcoming Ceremony" (6 March 2001). Available: WWW URL: http://www.dol.gov/dol/_sec/public/media/speeches/000306chao.htm.

   26. Doug Henwood, "Info Fetishism," in *Resisting the Virtual Life: The Culture and Politics of Information*, ed. James Brook and Iain Boal (San Francisco: City Lights, 1995), 167.

27. Reich's fundamental thesis is elaborated more fully in *The Work of Nations: Preparing Ourselves for 21st Century Capitalism* (New York: Vintage, 1991).

28. ODM, 24.

29. Charles R. Walker, *Toward the Automatic Factory* (New Haven: Yale Univeristy Press, 1957), xix, cited in Marcuse, ODM, 25.

30. Henwood, 167. See also Richard Appelbaum, who argues that "Reich . . . envisions a postindustrial utopia populated entirely by capitalists and members of the professional-managerial class, symbolic analysts all. In Reich's 'can-do' world of people who use their minds to make things happen, there are no workers to be found—at least not in those nations that invest sufficiently in their most valuable resource, their people." Richard P. Appelbaum, "Multi-culturalism and Flexibility: Some New Directions in Global Capitalism," in *Mapping Multiculturalism,* ed. Avery F. Gordon and Christopher Newfield, (Minneapolis: University of Minnesota Press, 1996), 302.

31. Marcuse, ODM, 27.

32. Karl Marx, *Grundrisse,* trans. Martin Nicolaus (New York: Penguin, 1993), 690–706. The "Fragment" is actually made up of two fragments from what was to be "The Chapter on Capital." For modern interpretations of the "Fragment," see especially Antonio Negri, *Marx Beyond Marx: Lessons on the Grundrisse,* trans. Harry Cleaver, Michael Ryan, and Maurizio Viano (Brooklyn: Autonomedia, 1991), 139–50; and Paolo Virno, "Notes on the 'General Intellect'," trans. Cesare Casarino, *Polygraph: An International Journal of Culture and Politics* 6/7 (1993): 32–38.

33. Marx, *Grundrisse,* 706.

34. Mike Davis offers the following anecdote to describe the underbelly of the information age, at least as manifest in Southern California's city of Vernon. I cite it here for its remarkable contrast to Reich's anecdotes above: "We are only about five miles from Downtown Los Angeles, but have entered a world invisible to its culture pundits, the 'empty quarter' of its tourist guides. This is L.A.'s old industrial heartland—the Southeast. It's 4:30 P.M. Two workers are standing behind an immense metal table, partly shaded by a ragged beach umbrella. A portable radio is blasting rock-n-roll *en español,* hot from Mexico City. Each man is armed with a Phillips screwdriver, a pliers, and a ball peen hammer. Eduardo, the taller, is from Guanajuato in north-central Mexico and he is wearing the camouflage-green 'Border Patrol' baseball cap favored by so many of Los Angeles's illegal immigrants . . . . Towering in front of them is a twenty-foot-high mound of dead and discarded computer technology: obsolete word processors, damaged printers, virus-infected micros, last decade's state of the art. The Sisyphusean task of Eduardo and Miguel is to smash up everything in order to salvage a few components that will be sent to England for the recovery of their gold content. Being a computer breaker is a monotonous $5.25-an-hour job in the black economy. There are no benefits, or taxes, just cash in a plain envelope every Friday. Miguel is about to deliver a massive blow to the VDT of a Macintosh, when I ask him why he came to Los Angeles. His hammer hesitates for a second,

then he smiles and answers, 'Because I wanted to work in your high technology economy.' I wince as the hammer falls. The Macintosh implodes." Mike Davis, "The Empty Quarter," in David Reid, *Sex, Death, and God in L.A.* (Berkeley: University of California Press, 1994), 54–55.

35. Roszak, 28.

36. See Marc Cooper, "Class.war@silicon.valley," *The Nation*, 27 May 1996, 12. Of course, while more recent developments in the high technology economy have had a significant impact on the class structure of the valley, the fundamental division of labor has not changed. See Ethan Smith, "Labor Pains for the Internet Economy," *The Standard*, 23 October 2000. Available: WWW URL: http://thestandard.com/article/display/0,1151,19555,00.html.

37. Everett Rogers and Judith Larsen, *Silicon Valley Fever: The Growth of High Technology Culture* (New York: Basic Books, 1984), cited in Roszak, 28. Chris Benner unpacked the full significance of this economic restructuring in a letter to the editor of *Wired* magazine: "First, while the number of knowledge-intensive jobs has exploded in the region in the last 15 years, so has the number of low-skilled, low-wage jobs. A large amount of low-wage computer and electronic assembly work is still done in this country, largely by Asian women working under difficult conditions. . . . The additional mushrooming of suburban development and shopping malls connected with the economic growth of the region has led to the expansion of low-wage retail sales and consumer services. Here in Silicon Valley, the heart of the information economy, the occupations with the greatest job growth include such 'low-knowledge' jobs as retail salespeople, receptionists, cashiers, and janitors. . . . Second, the rise in various forms of contingent employment—temporary, part-time, contract labor—means that even highly skilled workers face economic insecurity and downward pressures on wages. In Santa Clara County, the heart of Silicon Valley, contingent employment has accounted for *all net job growth* since 1984. This reflects the corporate trend to decrease the size of the core workforce and turn everyone else into economic shock absorbers of the highly competitive and volatile global economy. This creates problems for skilled and unskilled workers alike. None of these people has job security, and access to benefits and health care is difficult. Real wages for all occupations in temporary agencies have declined by nearly 15 percent over the last five years, according to the Bureau of Labor Statistics, while real wages for technical occupations have declined nearly 28 percent." Chris Benner, "Debating the New Economy," Letter to the Editor, *Wired*, June 1996, 30, 34.

38. Jerry J. Jasinowski, "In Defense of Big (Not Bad) Business: We Aren't 'Corporate Killers,' We're the Envy of Every Country," *Washington Post*, 17 March 1996, C1.

39. Scott Handleman, an intern at the *Nation*, pointed this out after he saw the $93,000 figure and immediately called Cypress Semiconductor to apply for a job as a receptionist. See Alexander Cockburn, "How to be a Receptionist on $93,000 a Year," *The Nation*, 8 April 1996, 9.

40. The $93,000 figure was originally claimed by T. J. Rodgers, the C.E.O. of Cypress, and reported in the *Wall Street Journal*. Rodgers, known to the

industry as the "bad boy of Silicon Valley," more recently gained notoriety in the corporate assault on "diversity" through his "blistering assault" on a concerned nun: "Bluntly stated, a 'woman's view' on how to run our semiconductor company does not help us," quoted in Ellen Joan Pollock, "Angry Mail: CEO Takes On a Nun in a Crusade Against 'Political Correctness'," *Wall Street Journal,* 15 July 1996, A1. Incidentally, Rodgers's heavy-handed rebuff to Sister Doris Gormley's polite letter won him huge accolades from the corporate community. In any case, while Rodgers's example is perhaps extreme, a certain measure of systematic distortion, as well as the simple and easily refutable lie, is more than commonplace.

41. As Spivak points out, "[h]uman labor is not, of course, intrinsically 'cheap' or 'expensive.' An absence of labor laws (or a discriminatory enforcement of them), a totalitarian state . . . and minimal subsistence requirements on the part of the worker will ensure it." Gayatri Chakravorty Spivak, "Can the Subaltern Speak?" in *Marxism and the Interpretation of Culture,* ed. Cary Nelson and Lawrence Grossberg (Urbana: University of Illinois Press, 1988) 288.

42. Marx, *Grundrisse,* 693.

43. The phrase "the ghost in the machine" comes from Gilbert Ryle; while it is an apt metaphor for the discussion here I should add that my use of it is completely different than Ryle's; in his work the phrase summarizes his critique of Descartes's mind-body dualism. See Gilbert Ryle, *The Concept of Mind* (London: Hutchinson, 1949), 31. I am grateful to Randolph Cooper for pointing to this text in a post to the electronic mailing list of the Computer Professionals for Social Responsibility, archived at http://www.reach.com/matrix/community-memory.html.

44. ODM, 32.

45. Jeans advertisement seen on a Sunset Avenue billboard, Hollywood, California, circa 1996.

46. Roszak, 19.

47. ODM, 23.

48. Ibid., 256.

49. Michel Foucault, *History of Sexuality Volume I: An Introduction* (New York: Random House, 1978), 95–96.

50. Ibid., 159.

4

On Globalization, Technology, and the New Justice

Tom Darby

oday the word *globalization* is on the lips of most people living everywhere on our planet. But this word never was uttered much more than a decade ago. Many have tried to accurately say when the term erupted into common usage, but on this subject there is no agreement. However, it is clear that the word would have made no sense before the world ceased to be multipolar and then bipolar, specifically before the end of the World Wars—WWI, WWII, and WWIII—the last of which we still insist on calling the "Cold War." While the world wars were about planetary rule, in that they were about who had the best means to the end of ruling the planet, they too were European (ergo, Western) wars, or one might say they comprised a century of European (and by extension North American, Russian, and Australian) civil war that brought the whole globe under its sway. But now that the century of world war is over and the planet is under the sway of a hegemonic West, it is possible to speak of globalization as a concrete, serious, and hence real phenomenon.

On Globalization

So, what a few years ago was esoteric and abstract now has come to be part of life for an increasing number of people on this planet. Our

experience of this shrinking world and our expanding picture of it, like
the very breath of our lives, in a mere few years has become a kaleidoscope of the real as imagined and the imagined as real. It is a world in
which the non-West is progressively transformed into versions of the West,
and the West shaped by that very Other it transforms. But this world,
which, not so long ago would have been unimaginable, deludes us. It
deludes us because it increasingly embraces us, smothering the mysterious
under the cloak of the everyday, denying us the experience of wonder. In
progressively becoming our common, virtual, yet empirical world, it also
is becoming a more vulgar world, for in this world everyone is either
already fit into the frame of the "Western Picture," clamoring to be within
its frame, or with increasing vigor resisting this new phenomenon.

I wish to begin with the most familiar, with those who resist globalization in the West; later, in the conclusion, I will compare them to
non-Western protesters. Western protesters often regard globalization as
the plan of some kind of "sinister inclusiveness" forced upon us all by
some shadowy group that "controls" globalization, a force often called,
or at least identified with so-called "corporatism." Here, at best we have
ironies, at worst, mere conspiracy theories.

It is advisable to ignore the conspiracy theories and to focus on the more
worthy ironies. The greatest irony is that the very Westerners resisting this
"inclusiveness" bound to and inseparable from the universalizing and homogenizing character of globalization—in their clinging to the categories we once
used to explain experiences that no longer apply to our world—namely the
categories of "Left" and "Right"—are constitutive of the most exclusive and
thereby the most non-"progressive" of political "movements" today.

What most, if not all, of these Western protest groups of globalization
have in common is not so much their seeing corporatism as the cause or a
cause of globalization, but in their assumption that globalization can be
authored, managed, or controlled. With this is the view that technology is
value neutral, that it merely is applied science. But this is precisely the view
held also by the National Rifle Association (NRA), the "poster boy" of the
most powerful "right"-wing lobbies in Washington, and their perspective,
captured by their mantra: "Guns don't kill people, people kill people."

The existence of such ironies can be accounted for, at least in part,
by the sheer disorientation brought about by the changes associated with
the modern age in general and with our late modernity in particular. Our
time really is different from previous ages of Western history. And this
is clear today, for it is only in our time that one can have a picture of
the world as a whole, albeit from every perspective.

How can this be so, and be so from every perspective? Perspective
has to do with where one stands as one views an object—the object,

in this instance, being our globe, our planet—and the point is that today we have an unlimited perspective from which to view the same object. This is so because today it is possible to view the planet from all standpoints, to have an unlimited "standpoint," hence a standpoint without standpoint.

What makes this new perspective possible is our technology. In other words, our technology is the independent variable of globalization. Or, put another and perhaps clearer way, neither globalization nor anything like it that could give us this view of the whole would be possible without our technology.

This is not to say that related phenomena—spiritual, political, and technical (but not technological)—that gave rise to the ironies and disorientation of today have not existed in the West and in other civilizations and in other ages. On the contrary, the spiritual crisis that led to the warring cities of Greece during Socrates' day is directly related to the technical (but not technological) ability of the Athenians to become an imperial, commercial, and naval power inciting the envy and hatred of her neighbors. The picture that Plato sketches at the beginning of his *Republic* is this very picture. But the world of Athens and of Greece was much smaller than ours. Yet it was a world or a "whole"—a cosmos— a city with clear boundaries, until her boundaries expanded and Athenian power spilled out into the space of the larger eastern Mediterranean. Later, the eclipse the Roman world led to the world of Christendom, which in turn, fell under the sway of modernity and its defining perspectives and practices that we call technology.

However, because all of these worlds are bound by their limited perspectives, the word *world* does not pertain in the literal way as we now speak of it (world = globe = planet Earth). Hence we use the word *globalization* because we *can,* and we can because it is the world that we now inhabit and that we can see. And what we see is this new perspective and its attendant practices that we have described as the phenomenon of globalization. However, it was not the power shifts of the last century that have given us the ability to view the world as a whole; underlying both our new perspective and those eclipsed political configurations is technology. *Technology is the independent variable of globalization.* Hence, any attempt to understand this new phenomenon without first understanding technology is a waste of time.

On Technology

So, what is technology? One can begin with this thoroughly modern word and try to recover its meaning synthetically. *Technology* is a compound of the two Greek words *techne* and *logos*. *Techne* pertains to the universalizing and homogenizing of making *(poesis)*. This re-presentation in the form of a thing (= object = a being that stands before us) is demarcated by clear boundaries (= completeness = wholeness). This is why we can say that a "thing," in its constituting completeness is an ordered whole that is fit for a purpose, or, therefore, is an End in itself. Hence, making is about Ends, in the sense that the finished object or thing justifies itself.

Making, while a necessary condition for the appearance of technology, is not sufficient for the appearance of technology. This so because technology is not a compound in word only, for it is compounded from the co-penetration of making and knowing. *Logos,* in archaic form, is a gathering together of experience that leads to perception, or, we might say, to knowledge. This is why the "word" *logos* often is translated as "word," in that a "word" is the result of the gathering together of experiences culminating in knowledge.

But *logos* does not refer just to any gathering of experience, for, because knowledge is its result, *logos* refers to a way of seeing. This is why it often is said that *logos* is a "way" or even "the way." To speak of a way is to speak of means (as opposed to ends). So, technology in pertaining to t*echne* and *logos,* at the same time, is about means and ends.

Technology is the progressively rational (= efficient) arrangement of means and ends (for humans) and cause and effect (for nature). The former, therefore, has to do with possibility, that is, perception, and the latter with actualization, hence, with practice. Technology has as its project the transformation of nonhuman and human nature.

By "nature" I refer to the limits and possibilities of any being. Nature, therefore, is whatever exists and is not otherwise. Nature *is* that which appears to be *what* is. It is the "whatness" that appears to us. For this reason nature both nonhuman and human often is called the "given," the given referring to those boundaries that make it possible for us to compare and to contrast one being with another, thereby allowing for our ability to measure limits and possibilities, which, in turn, enables us to classify, and hence to name.

Measuring, classifying, and naming are all aspects of *justifying. Dik*e, the Greek word for justice in its most archaic form, pertained to a pile of stones that marked the boundary between a public road and a sacred grove, hence to the demarcation between the spheres of sacred and profane. The point here is that justice has to do with the boundary itself. To be exact, justice is the in-between, a magical threshold that allows one to see the inside and outside, the up and the down, all at once. So, justice also has to do with seeing and therefore with the kind of knowledge necessary for knowing what is or is not fit for one to say or do. In the example of the sacred grove along the road, being able to discern the boundary is a prerequisite for our knowing if in crossing the line we are doing what we are fitted to do—that is doing what is right (= that for which we are fitted = just) or trespassing.

However, there is a difference between the ancient traveler and ourselves. For the ancient the pile of stones marks a *given* boundary, but for us there are no given boundaries that cannot be altered or obliterated; hence, there are no given standards that are inviolate. Nature and justice as standards for true speech or right action are at the heart of what we call technology and we will see why technology dictates them problematic.

While nature and justice are not themselves standards of the modern age this does not mean that the modern age has no standards by which to measure fit speech (truth) and fit action (right), that is, justice. The standard for modernity is technology, and the standard for technology is efficiency. Efficiency, the end (= purpose = completeness) of the projection of technology constitutes the boundaries of technology. Efficiency can be measured only as a progressively diminishing difference between these means and ends or causes and effects. Thus, technology is 1) self-referential, 2) relatively autonomous, and 3) progressively sovereign, and, being so, 4) tends toward the systemization of nature both human and nonhuman. If the relative difference of means and ends (or cause and effect) were ever reduced to zero, or to complete efficiency, then technology would become a totality (i.e., a total or complete system).[1]

Although in our time technology is embraced by not only the West but by the non-West as well, technology is a compound of Western perceptions and practices. The perceptions point to the radically revised relations of God, Nature, and Man that crystallize in early modernity, but go farther back to practices of radically increasing prowess,[2] which, in turn, dynamically shape, and are shaped by those radically revised perceptions. First, I refer to the perceptions of Bacon, who urged us to put nature on the rack and to vex and torture it so as to force its "reasons for being," and to Hobbes who told us that man's artificial creation, the "World," is superior to God's natural creation—the Earth and the beings upon it—except, of course, man. Bacon was among those whose observations of nature led to what we call "the scientific method," which, in turn, enhanced our control over nature qua nature, through developing a way of transforming nature, and eventually systematizing it.

But, as Martin Heidegger notes, Bacon's new perception of nature was not enough, in that it pertained exclusively to perception. What was needed to bring about the scientific method was a practice coupled with this changed perception. This practice—a practice that allowed man to act upon nature—resided in imitating God's action, in creating the world and His eruption into the Earth and the World, that resulted in His embodiment within the realm of nature (Space) and history (Time). I refer to what lies at the heart of Christianity itself, to the doctrine of the incarnation of Christ. This doctrine became a practice with the medieval Schoolmen who attempted, through their interpretation of the Word, to embody the divine Will itself. But I suspect that these roots lie even further back, in the cabalistic and gnostic practices of both Jews and early Christians.[3]

The perceptions of men such as Bacon, together with those practices rooted in Christianity, result in attempts to systematize nature qua nature, while Hobbes systematizes human nature. Hobbes begins with his attack on God's creation and Aristotle's doctrine of causality. His "artificial man," Leviathan, is a systemization of human nature, in that it is a system built by man—man as the Maker who makes Himself.[4] But this New Man is a creature with neither conscience nor a longing for transcendence. He has exchanged his conscience for the rational calculation of self-interest and his human longings for transcendence for his immanent safety. Bacon and Hobbes were among the men who discovered that the power of modern science (= technology) lay in its tendency toward systemization. Because of this, they are harbingers of the new age for the West.

But, according to Heidegger, this still is not enough, for in order for it to be, technology requires the advent of *research,* for without research, there is no procedure, or way *(lo modo),* as Machiavelli calls it (lo mode = the way = the way of of today, hence what is the modern; e.g., a present oriented toward the future). However, research is not just procedure, it is the projection into nature (into what *is*) of that "fixed ground plan," or, perhaps better, what is rendered as "strategy," *(Grundriss).* The projection draws (wills) the boundaries, in advance, and the way of knowing must adhere to these *orders* or boundaries. Heidegger calls this "binding adherence" to research "rigor." Projection of the fixed ground plan is the *first command* of research. "Science becomes research through the projected plan, and through securing that plan in the rigor of procedure."[5]

Its *second command* is methodology. Methodology is the way of clarifying the known, and relating the unknown to it, thereby increasing the sphere of the known as facts. This is at the heart of what I call metaphor (*meta* + *phoros* = to speak across = to connect different experiences or perceptions). This leads to explanation, explanation to law, and law to experiment; the latter itself, mirroring—albeit in a disembodied or abstract way—learning itself: "The more exactly the ground plan is projected, the more exact becomes the possibility of experiment."[6] Exactness leads to the objective knowledge we call "facts," that is, information (= decontextualized knowledge) because the "ground plan" that is willed and projected is both controlled *before* the experiment itself, yet continually adjusts itself to its results. Thus, the way of method is what I identified above as the self-referential, self-adjusting aspect of technology.

The *third command* of research is that it be what Heidegger calls "ongoing" *[Betrieb]*, that its activity both pertain to its proper (ordered = bound = fixed) sphere, that it, in other words, be specialized, and that the facts be coordinated so that the methodology can be adjusted to the results. This simply means that one must specialize and that specialists need to communicate and cooperate. This is why the business of research must be ongoing.[7] "Ongoingness" (my own ugly Heideggeresque term), as I see it, is the researcher's unswerving attention to the command of technology and the willingness to follow its command.

Research brought technology to this point. But because of what became an overwhelming mass of facts (again, facts = information = decontextualized knowledge) generated by it, a new way of ordering, storing, and explaining was necessary.

That which concerns us most as members of a family, tribe, village, town, city, or larger cultural or civilizational unit is what I call the "Underlying Concern." The way we make sense of these experiences that most concerns us is what I call the "Overarching Metaphor." Each "Age" is defined by the relation of its Underlying Concern and its Overarching Metaphor.

For example, in the history of Western perceptions and practices, there have been several configurations. First, because of the radical changes that culminated in the spiritual crisis caused by Athens' aforementioned loss of her war with Sparta, Plato saw that the Overarching Metaphor of the Greek polytheistic pantheon could no longer explain Greek life. He saw that the Overarching Metaphor had become bereft of meaning and hence had lost its authority. Plato related the "new" experiences of the Greeks to the patterned change in the heavens shared by all Greeks, and upon this relation, philosophy was founded. Aristotle too used patterned change or "nature" as his metaphor, but rather than relating the experience of the citizen to the city and then the patterns in the sky, he found his patterned change in the biological world expressed as the relation of species to genus.

The Hebrews' metaphor was their relation to their God, manifest as their covenant with Him, their history as an enactment of His will. St. Augustine in his blending together these Hellenic and Hebrew metaphors gives us the City of God and its relation to the city of man, the explanatory power behind Christianity. This metaphor—albeit with much difficulty—served to explain experience to the Christian West from its inception during the fifth century C.E. until the modern world.

Due to the radical and rapid changes compounded into the perceptions and practices (ways of viewing and ways of doing; hence, ways of thinking and ways of acting) that constitute technology, the modern world has had three phases, or what elsewhere I have called "waves."[8]

Each wave has had its own metaphor rooted in its own experience and its own symbol that best allow for ordering, storing, and explaining each phase of the experience of modernity. Each metaphor, along with its symbol, has been technological.

The experiences of Bacon's and Hobbes's day required a *mechanical metaphor*, which was encapsulated within the symbol of the clock (= kinetic system). Because symbols are wholes, they too are systems, as is the very machine we call the mechanical clock. The day of Kant, Hegel, Marx—and to a lesser extent of Nietzsche—needed a *hydraulic metaphor*, symbolized by the gun/engine (= the digestive system).

During the previous century, and for most of the twentieth century, the hydraulic metaphor, and its symbol, the engine, have sufficed, and indeed, most people are still stuck with the vocabulary based upon these older metaphors and symbols, in that we still describe both nature and human nature in terms of forces, pressures, processes, and movements. Our everyday assumptions about both nonhuman and human nature are those mechanical perceptions granted by Galileo and Newton, and the hydraulic perceptions associated with the second law of thermodynamics.

While most people today (including philosophers and scientists) still rely on these metaphors and symbols for explaining their experience, reflecting on Heidegger, I think that while all technology pertained to a summoning forth of energy from nature, transforming it, and storing it for future use, that something fundamental had changed. Heidegger calls this process of extraction and transformation "Enframing" *(das Ge-stell)*, and its storage for future use, "standing reserve."[9] Although still in embryonic form, what Heidegger saw as so basically different was that energy could be extracted, transformed, and stored differently, and that which, in his part of this century, was hardly manifest, at the beginning of this century is commonplace.

Having said this, let us now take an excursion through the three metaphorical attempts to explain what has concerned moderns most in our technological age, an age beginning somewhere around 1600. Modernity, I remind you—because of rapid and hence disorienting change—has had several overarching metaphors, all of them technological. We have had a sketch so far of two. So, we go from there. And now we must go farther down to the deeper fundamentals.

First, mechanical energy was extracted from nature and stored in the weights and springs of machines such as clocks—machines that held the energy of nature in reserve until they transformed it by setting it on its way, by unwinding a spring or releasing a weight. Next, hydraulic energy was extracted from nature, transformed, and reserved within the walls of a frame called an engine and set on its way by releasing water, steam, or a regulated explosion to drive a turbine, a piston, a jet engine, a rocket engine, or, at the end of this wave, crude nuclear power. But what fundamentally has changed is that now we are able to extract, transform, store, and set on their way the less apparent, even invisible, yet fundamental energies of nature. I refer to what lies within our inner space such as the DNA of our bodies and to that energy within the atomic structure of all bodies (beings), electricity—the spark of life, and, one might say, its *spirit*. I speak of the wave of the *electric metaphor,* the symbol of which is the electric computer, without which the present phase of the Western project, the "end of history" and "globalization" would not have become possible, much less apparent.

The universal use of the electrical computer marks the *appearance* of the coming together of perception (knowing) and practice (making) through the co-penetration of the computer's superstructure (its software = perception) with its infrastructure (its hardware = practice). I say *appearance,* as this coming together only appears to be, in that the space between them is relatively but progressively invisible. This is so because *in* the space in-between is time itself. Granted, we see neither force nor pressure, but we see their effects, relate them to their causes, and call this change. But the change we see in our time is coming so swiftly that we see it less and less, and we are progressively coming to consider what we don't see as normal, for the normal is precisely what is nearest, and therefore, not questioned. Time is Being. And both Time and Being are progressively becoming invisible.

Whereas, the power *(efficiency)* of mechanism is measured as force, and that of hydraulicism as pressure, the power of the computer lies in the difference in-between the on and off pulse of a charge of electricity, and thus is measured by *speed*. Hence, the technology of mechanism and hydraulicism is manifest in the representation *(Vorstellen)* of the apparent as objective, ergo concrete, in that it demands the centralization, hence the massification of force and pressure in machines; and in human technologies, the massification of money in economies, the massification of people in societies of large nations in great cities, governed by extensive bureaucracies, all protected by great armies. So the power of the technology of our day is derived from the relative but progressive rate of diminished time, with the apparent disappearance of the representation of the time that the West has called history.

Thus, electronic technology demands not the overstatement of appearance (representation) that in the time of the mechanical and hydraulic metaphor Heidegger calls the "gigantic,"[10] but the dissolution of the boundaries in which power was previously contained, hence the decentralization and dispersion of power. Artists, as usual, realized this first. Witness the blurring boundaries of the image in impressionist art, then broken images in cubist, multiperspectivist (hyper-relativist) art and finally, the universal homogeneity of abstract art. Now, there is our present antireality, and hence reactionary movement—so called postmodernism. While artists need not account for what they see and do, unlike those modernists, most postmodernists are not artists, nor are they philosophers. Rather than questioning what *is,* they resent it, and so, with their rhetoric, try to conjure it away, but in their reaction to modernity, are, in turn, conditioned by it, and hence, unwittingly, are an integral part of it. Postmodernism is the last puff of the age of the engine: history's fart.

But in terms of the serious demands of our technology today, I am thinking of the level of coordination in scientific and humanistic research, and how without the advent of the computer this would not be possible. I am thinking of electronic communication in general, but specifically of television and the "World-Wide Web," and how these two modes of communication are destined to come together and then merge with technologies we now are unable to imagine. And, I am thinking of how all of this makes us witness to the ever swiftly disorienting eclipse of the sovereignty of the nation-state as it thrashes about to control these technologies through legislation justified by pietistic rhetoric.

As we witness this, a new political actor is waiting in the wings, ready to present himself on the world stage. This new form of rule—this new way of seeing things and doing things—has a shape too vague to define. We who live today at the eclipse of the second technological metaphor, the Age of the Engine, and in the dawn of the age of the third technological metaphor, are, rather, witnessing what Hegel describes in his *Phenomenology* as the "New World." This political form which Hegel described turned out to be the beginning of the universalization of the nation-state, and in turn, the arrival of the conditions for both its demise and the hazy outline of the political form that will eclipse it, bringing about the Age of the Computer. But because we cannot yet name this new political form we simply are left with describing it in relation to the other two technological metaphors, the mechanical and the hydraulic.

Our technology at the beginning of this century is manifest by the disembodiment of power in the form of the appearance of the invisible. This is why the power of technology now often is referred to as *soft power,* soft because it is both malleable and boundless. It is a technology the power of which either appears benign or, because of its stealth, appears not to exist at all. Its use always is justified by that abstraction called "values." This power is as soft and as elusive as the electronic image itself. Mass communications is both centralized and dispersed power. It also is mass illusion and delusion in that the more decentralized and dispersed it becomes, the less natural and historical reality exist.

Given the "world picture," more and more people are coming to take the virtual as an improvement over the givens of nature and history; or they simply are taking the virtual itself as the given, and therefore, not questioning the picture they see.

So we zap the TV. There is CNN. Well, maybe not. Perhaps a talk show in Icelandic, Slovak, or Urdu. But for now, most are in English, except, of course, Al Jazeera. Tomorrow? Maybe Chinese. Crudely put, news or talk about the news is "history as journalism,"[11] as Heidegger termed it. This is the world picture—the world as a picture—the world pictured as a whole in which the past and future are zapped into an electrical image of the present. The specific language matters less and less, for the format (the frame) increasingly is the same, for the content is conditioned by the context and the context by the perspective. "Truth is relative," so they say, but this is not the point. While truth is relative to the perspective from which the world is being viewed, limiting, thereby, both what one "sees" and how one interprets it, greater numbers of people are seeing versions of the same picture. Thus, more and more people are becoming less and less tied to their little corners of their necessarily limiting standpoints—and coming closer to what Heidegger calls a "standpoint without standpoint."[12] This point is so obvious that it likely is to be missed (and this is the point), for it is about that which defines us most, but which we question least. It is about our "Archimedean Point," our technology in general, and, electronic technology, in particular.

Let me repeat myself: technology is our common denominator, our independent variable. Thus, it both defines the world upon which we stand and our view of that world. Simply put, no thought of our world makes sense without taking into account the phenomenon of technology. Its major reason for being depends on our perception that all there *is* is only in relation to us, and thus, is there for our *use*. But because of technology, we are able to *do/make* what we *see*—to re-present a universe as we will to see fit.[13]

But representation entails negation, in that it is the given that is re-presented—both the historical and natural given. The magic of negation lies in its transformative power. While technology is about the transformation of the given, all that has been given to us is not yet transformed.

On the New Justice

One must conclude from the above that while one now can conceive of globalization, the planet has yet to be universally and homogeneously, ergo, completely transformed. Because of this fact we should remember that technology is about the relation of knowing and making—that is, viewing and doing. We have the perspective that allows us to view the whole (i.e., the planet), and while we may even go so far as to call this perspective of the whole (i.e., globalization) technology's "concrete universal," our means for concretizing this universal, that is, to make it actual, still is in process. Thus, globalization is a process that aims toward total efficiency and this is why, during this age of technology and globalization, efficiency is the only rule or measure, the only authority that we invoke. I say this because our old authorities that held fast to the metaphors that account for experience—a God of History, the Good of Nature—the authoritative standards for our positive law—have given sway to the measure of Efficiency. So we now must ask the question: Since efficiency demands a coordination of cause and effect, means and ends, hence a transformation of the given into the same, and since justice is a measure, in our increasingly boundless world, what kind of justice is Efficiency? This, I think, constitutes the reason behind our present and future conflicts. These conflicts are about the relation of globalization, technology, and justice.

In order to attempt to see even an outline of an answer to this question, we must now return to our present and to those who are violently protesting against this transformation of the planet that we now call globalization.

Globalization is just the latest, but so far, the most profound version of what we for a while have called "modernization." However, the relation between globalization and the modern are in dispute and the disagreement concerning the relation has to do with how one views technology and its relation to justice. I mentioned above the opposed groups of Westerners who oppose globalization, groups such as: environmentalists and farmers; white-power thugs and the North American Indian movement; Basque nationalists and self-proclaimed anarchists; the "slow-food movement" and European neo-fascists. All of these groups whose resistance to modernity is in the form of the anti-globalization movement are profoundly, yet usually unintentionally, connected to the non-Westerners who violently oppose globalization. I refer, of course, to the violent opposition unleashed on September 11, 2001. Let us look more closely at the Western and non-Western protesters.

Many of the Western protesters see themselves as heirs to the protest movements of the last century. The most apparent proof of this is in their recycling the styles of the 1960s and 70s. Yet, the difference is that at least most protests of the twentieth century sought changes that were seen by the protesters of that time to lead to justice, in the form of civil rights, for example. The protesters of the last century wanted change. However, neither the self-described "hippie anarchist" who sees the antichrist as a hamburger joint (McDonald's) or as a coffee house (Starbucks), nor the white-power thugs on the hunt for some dark-skinned immigrant who knows how to cook curry, want change. In fact, they are fearful of change. The Western protesters of globalization are profoundly conservative. Globalization is correctly seen as the agent of change, the change that has make it possible to have a Big Mac in Beijing or to enjoy what now is the favorite food of London: curry.

Globalization, for the anarchist hippie amounts to the grave injustice associated with what is seen as a global corporate takeover of culture, of nature, of—well—the planet—by those who "control" technology. And, for the white-power thug, globalization has written the ticket allowing the hoards of curry-making dark-skinned people to steal the rightful opportunities of white people. But the relation of globalization to justice not only connects these two quite different Western protesters, it too is what connects them to the non-Western globalization protester, the most extreme example being the terrorist whose actions are but a reflection of what he sees to be the grave injustices wrought by globalization, the gravest of which is seen to be destroying his religion and his traditional ways of life, while mocking his backwardness and weakness.[14]

Conclusion

It should be said that while neither the Western nor the non-Western protester understands the phenomenon protested, these protesters are correct in connecting globalization with justice. But both the Western and non-Western protesters are correct for the wrong reasons. However, when I say that both groups are correct in their making the connection, I am not forgetting that there are many different kinds of people from both the West and non-West who do not like this new justice, and the reasons for their protesting globalization are quite varied. However, although their reasons take several forms, none among them has any understanding of globalization. Their profound ignorance is due to their lack of understanding of the phenomenon and meaning of technology.

Were they to understand the "nature" and "meaning" of technology they would know that its only standard is the justice of efficiency.

First, if the protesters of globalization knew this they too would know that one cannot have modern (= technological) practice without modern (= technological) perception, and this, of course, dashes the hopes some have expressed for a technology bound by something so non-Western as Islamic law.

Next, they would know that globalization, in its being driven by technology, is a process that is *relatively autonomous* and *progressively sovereign,* and because of this, unlike the little man behind the curtain in the Wizard of Oz movie, there is no wizard and there is no curtain; yet Oz is real.

Last, they would know that once on the technological road to Oz there is no turning back, and getting off the road and striking out into the bush is unthinkable, for to get off the trail would bring unthinkable injustices, perhaps resulting in the destruction not only of the West but of the non-West as well. So, old answers do not satisfy new questions. In this new world we must find that for which we are fitted. It will likely be a long and hard search, and should we find it, it may not be an easy fit. Yet, we still can question, so we can still hope. And our hope is the proof that we still are human.

Notes

1. According to the second law of thermodynamics, systems tend toward totality, and totality leads to entropy. In relation to the emerging "system" of "planetary culture," entropy (= nihilism).

2. This radically increasing prowess appears early in the modern age with what I have called elsewhere, interrelated "Sovereign Regimes" of religion, art, science, and politics, appearing in this order. See "The Three Waves of Technology" in *The Literary Review Of Canada [LRC]* (Oct. 1995); and also see Heidegger who inspired my elaboration of his idea in *An Introduction to Metaphysics*, (New Haven: Yale University Press, 1959), 48. Incidentally, my three waves have nothing to do with Toffler, but everything to do with Plato. On Plato, I elaborate below. See n.8.

3. Heidegger, "The Age of the World Picture," in *The Question Concerning Technology and Other Essays* (New York: Harper Colophon, 1977), 122. My suspicions are best explained by "Epilogue: Gnosticism, Existentialism, and Nihilism," in Hans Jonas, *The Gnostic Religion* (Boston: Beacon, 1963), 320. Also, see, Eric Voegelin, *Science Politics and Gnosticism* (Chicago: Henry Regnery, 1968); and Flannery O'Connor's novel, *WISE BLOOD* (New York: Alfred A. Knopf, 1962).

4. In his three-page introduction to *Leviathan*, Hobbes attacks Judeo-Christianity and philosophy. This attack is Hobbes's hypothesis or, the projection of his "fixed ground plan." Here he says man is superior to God, which is the same thing as saying there is no god but Man, and he metaphorically describes Man's "artificial creation," Leviathan, as a clock-like machine. In his attack on the heart of philosophy (science)—Aristotle's "Doctrine of Causality"—he compounds Aristotle's first and third causes (material and agency) and the second and forth causes (idea and purpose), thereby bringing together a new practice and a new perception. What follows, e.g., the text itself, is a *proof* of his ground plan. As Hobbes advises: "Read thyself."

5. Heidegger, "In the Age of the World Picture," 120.

6. Ibid., 122.

7. Ibid., 125. The "ongoing activity" of research requires institutionalization. In our universities the researcher will replace the scholar, as, indeed, today this almost has come to be. It seems that the only refuge for erudition left is in the liberal arts *(artes liberales)*, a small enclave for humanity, perhaps its last hope for dignity during "the age of the world picture."

8. Darby, see n.2, above. While in actuality globalization is new, the experiences engendered by it are old. These experiences are the theme of *Republic*. In Plato's dystopian book V, Plato, playing on Aristophanes' "The Assembly of Women," outrageously eradicates the difference between the public and private realms, by 1) eliminating eroticism, then 2) the family, and last 3) the difference between action and thought. These conditions are his famous "waves." His first wave is about *universalization,* his second about *homogenation,* and the third is the end of politics and philosophy with the reign *(state)* of the philosopher-king. This is the greatest of Plato's serious jokes. And, then there is Genesis 11: 1–9, that recapitulates the Fall in the story of the Tower of Babel. But then there was God and the Good. Destiny has a beginning and an end.

9. Heidegger, "The Turning," in *The Question Concerning Technology*, has Enframing and Standing Reserve as its major theme. My term for "Standing Reserve," has its shallow roots in our late-modern experience and is part of our worn-out late-modern vocabulary. My word for it is simply, "Stuff." See my *Sojourns in the New World: Reflections on Our Technology* (Ottawa: Carleton University Press, 1986).

10. Heidegger, "In the Age of the World Picture," 153. An explanation as to why the U.S.S.R. fell: it was not able to make the transition from "hydraulics" to "electricity," and instead of exploding (= revolution), merely imploded from leaking pressure due to a lack of fuel (= money), thereby collapsing under the gigantic weight of its engine (= frame). Now the rubble (Russia) has reverted

to a frontier (= lawless = criminal) society somewhere between "mechanism" and "hydraulicism." The U.S.S.R. tried to escape the past and today Russia is being squashed by its past as the U.S.S.R., and a series of anemic revolutions could pop out anywhere and anytime. Yet, the Russians still make the best rockets, e.g., engines, their only significant contribution to the International Space Station Project. For now, the micro-electronics of the "electric age" still must depend on engines to move its hardware, e.g., its frame (= body = being). Tomorrow, when another way is found, perhaps the Russians will become like the early-modern Chinese, who, after having invented gunpower and ballistics, then were reduced to making and exploding firecrackers to ward off the spirits of evil ancestors.

11. Heidegger. What is meant by "History as Journalism" is the following: 1) politics and its result, history, is transformed into culture and culture into entertainment, 2) imploding the past and future into the present, and 3) instant global communication. See, as described in *An Introduction to Metaphysics*, 37–38. During the last days of the last century we were entertained by the electronic implosion of the drama of sex and death on CNN—the impeachment of the American president during the electronically controlled bombardment of Iraq. Also, there was the TV horror show in real time. Yes, I refer to the events of September 11, 2001.

12. Heidegger, *Nietzsche: The Eternal Return of the Same*, Vol. II, trans. David Farrell Krell (New York: Harper and Row, 1984), 117.

13. Heidegger, "In the Age of the World Picture," 132. Also, I remind the reader that while CNN carries this picture, so does Al Jazeera.

14. For the best short yet convincing account of why the Islamic "world" is clashing with the West see Bernard Lewis, *What Went Wrong? Western Impact and Middle Eastern Response* (Toronto: Oxford University Press, 2002).

5

What Globalization
Do We Want?

Don Ihde

The title captures the double sense that indicates that globalization carries with it both an element of choice and yet also non-choice. What is needed is to orient ourselves. The focus here is the technologies that enable globalization and the forms of embodiment these technologies entail. First, let us take a look at some technologies that relate to the international, communicational, and transportational dimensions of globalization.

Drawing from common academic experience and casting the narrative in a phenomenological framework, how is globalization experienced, *perceived*? What does globalization look like close up? And, correlatively, what requirements are there for globalization to look like whatever it turns out to be?

For example, here is a recent e-mail request, probably not unlike many that other academicians receive: "I am now preparing an article on the issues of globalization and digital art (entitled: World Wide What?). Finn Olesen (Media professor in Aarhus, Denmark) has told me that you have written about globalization as a phenomenon that took its very beginning when we had the first photography of the globe. The whole idea of seeing the planet Earth as a whole for the first time grounded this new concept. Needless to say, from my photographic perspective I find this approach extremely interesting and a groundbreaking way of thinking visual representation. However, Finn could not remember where you had written about this issue. So I was wondering if you could possibly refer me to the relevant article or book?"

Then, my response, which recognized the reference from a 1996 article on "Whole Earth Measurements," with the subtitle, "How many phenomenologists does it take to detect a Greenhouse Effect?" The article takes up questions and technologies related to how one could measure possible earth warming. And, relatedly, there is the question about how much of a warming effect is "homogenic," or related to human activity. The very notion of a whole earth measurement is distinctly late twentieth century. I was arguing that for these ideas to make sense, one has to have some notion about how science and technology are a material *technoscience,* but also one has to have some way of dealing with the entire *Earth-as-planet:*

> Classical phenomenology, I argue, lacks the first of these concepts, and Heideggerian romanticism rejects the latter. Let us begin with a very simple phenomenological question: from what standpoint or perspective can the issue of whole Earth measurements be made? If, at base, our very knowledge is constituted by way of our *bodies* and through *perception* as both Husserl and Merleau-Ponty seem to contend, then what is claimed about whole earth measurements becomes problematical . . .[1]

In the context of the original article, I then go on to show that *only through instruments do sub-perceptual entities get perceived and that this instrumentation is essential to technoscience.* I conclude that it is Husserl's concept of science that is deficient since he concentrates—as all early-twentieth-century philosophy of science did—upon science as primarily a phenomenon of mathematization rather than instrumental, technological embodiment, yielding new perceptions. Husserl's Galileo remains a Galileo without a telescope! A different critique applies to Heidegger who conceives of Earth as ground and that in the context of extreme romanticism about nonmodern and nontechnological notions. But for purposes here, I am concerned with how to get a perspective on whole earth phenomena:

> The . . . conception needed is *perspectival* in the sense that one must have a sense of a "whole Earth" as that field which is measureable and this is what I am calling *Earth-as-planet.*

The idea of Earth as planet, as a finite sphere, is itself quite ancient: it is anticipated by Aristarchus, it is assumed by Copernicus, and its size has been measured along with these notions, both in Greek and early Modern times. But one could argue that it is not *seen* as planet until it is so embodied in the earth shots with which we are now all familiar. And it is here that I enter the last set of arguments with classical phenomenology. . . . Put crassly, what might Husserl and Heidegger see when they look at a Moon shot of Earth-as-planet, or when they look at other imagings of environmental phenomena? If the answers are "pictures" or "images," then the look involved is both naïve and cast in terms of modern epistemologies which remain caught with a passive theory of perception, which sees only "surfaces" and "representations." And, if in the Heideggerian case, there is more penetration, but a penetration which yields the referent phenomenon as itself a "picture," then the seeing remains short of *scientific seeing* as evident in science praxis. Indeed, such a look would remain short of even the primary insights of a phenomeno- logical epistemology claimed by both but restricted to every- day or prescientific lifeworld regions.[2]

Although I have here "cut and pasted" an exchange from commu- nication and publication technology, the above illustration reveals an- other set of technologies that relate to another global phenomenon concerning the environment. It should be clear that any whole earth phenomenon also entails some dimension of globalization. And that problem as a *global* problem is emergent as a twentieth-century phenom- enon. That is, to have elevated to an explicit theme the environmental "health" of the earth, is different from the prehistory of environmental- ism, which contained differing attitudes toward nature, animal life, or extra-human life that were embedded in different religious and cultural beliefs and practices. Thus, the analogy between the abstract knowl- edge that the earth was a sphere, finite, and in motion within the heavens—knowledge gained in antiquity—and that same knowledge perceptually experienced, either at first hand by astronauts, or mediated through photography, as in NASA earth shots. To have become a rec- ognizable part of a *lifeworld* implies this much more recent perspectival perceivability. To that extent, my Danish compatriots had a point con- cerning claims about globalization. And, to that extent, if one has a concern for the global environment, one must include this dimension in any wanted globalization.

But let us now take a second slice through this opening to the experience of globalization. The e-mail correspondent, Lone Sonderby, and Finn Olesen, are Danes. The concrete occasion from which this exchange occurred goes back to a graduate Internordic Seminar in Aarhus, Denmark, last March; communication about this seminar continues "virtually" through e-mail. An academic lifeworld is today itself "global" and is embodied in a series of alternations of face-to-face and electronically mediated communications. In one sense this alternation of face-to-face and electronically mediated meetings is merely a variant upon a much older form of face/virtual communications, now embodied in a different technology. Many years ago I had considered doing my dissertation on Georg Johan Hamann, an Enlightenment contemporary and friend of Immanuel Kant. Hamann was a philosopher of language and an essayist who knew most of the principals of the times and had an enormous correspondence with almost anyone who was anyone in the eighteenth century: Kant, Herder, and dozens of scholars of the time. But, instead of e-mail, he wrote by hand—not even the typewriter had yet been invented. And, re-reading this correspondence, we can no doubt marvel at the high art of literary expression undertaken by hand. Here, too, was an alternation of occasional face-to-face meetings (now lost to us since not recorded) and written correspondence. (Note an irony here: only the virtual, or technologically mediated communication remains!) Without casting any negative, or for that matter nostalgic and romantic evaluations on this face/virtual alternation, one can immediately see by variation how different today's version is from the eighteenth century's. Spatiotemporal pacing his drastically changed: letters by coach or sailing ship took weeks, months, or even longer to reach their destination. A written response, once received, would be thought through and slowly penned, and in the end it might be several scripted pages long. But, although by contemporary standards this is "molasses" space-time, it remained an alternation between face/virtual communications.

An example of such lifeworld alternation can be found in any short period of e-mail cataloguing: In the last few weeks I have communicated with Andy Feenberg, currently teaching in Kyoto, Bruno Latour, about to vacation in France, many of my colleagues in Stony Brook—from my summer home in Vermont; Donna Haraway, in her commune retreat in California, responded to the already cited Danes, and in addition helped with the visa processing for two more Danes coming to Stony Brook in the fall, thanked Laszlo Ropolyi in Budapest and Nikos Plevris in Athens for their arrangements for recent trips. Add now the frequent communications with graduate advisees, publishers, and editors, to which should be added numerous messages from many countries with inquiries and perhaps once-only correspondence. All of this is, if not "real time," at least occurring within the parameters of a few days. This set of exchanges is repeated in the experiences of many academics in any contemporary, interconnected pluri-university. It is a life textured by "global communications." This globality compresses older space-time into a kind of intimate immediacy, a near-distance whose space-times are equivalent for all geographical places. In a practical sense, I would find it very hard to have a style of life that entailed communication and travel without this compressed alternation of face/virtual communication.

This contemporary form of virtual/face-to-face alternation is both the same and different from older variations of the same phenomenon. It is the same in that it is an alternation between several types of communication and human interchange. And it is important to emphasize that the alternation itself is important—Jean Paul Sartre's drama, *No Exit*, in which the theme is that "hell is other people," expresses the possibilities of too much face-to-face, and for those in closed communities, for example monastic communities, it is perhaps inevitable that some such communities would adopt rules of silence! Nor would a virtual-only world be a happy one, and while I cannot really imagine such a world, if elevated to virtuality only, a communicative world of this sort would simply be the inverse of the enclosed tightness of the former.

The difference, of course, is that globalized alternations relate differently to both the spatiality and the temporality of those alternations. Potentially the virtual is *global* in extent, and experienced in a condensed *temporality*. The technologies that enable such extent and durational changes, however, are "soft" in the sense that we need not use them— I for example, do not have nor like cell phones, so I simply don't have one—or we can use them over any range of uses. I still know academic friends who refuse to do word processing, but also know compulsives who get sucked into hours of time "on line."

Having now begun with two common, related experiences of globalization, I now want to move outward and begin to approach the more normative dimensions of the question: "What globalization do we want?" Not everyone wants globalization, even if in some ways some kind of globalization is "inevitable" or is "already at hand." And, while there is a need to maintain a critical attitude toward the culture formations that will result from any kind of technologically mediated globalization, it is also wise to avoid two traps that many critics fall into.

The first trap may be called the "functional contradiction" trap. With only slight caricature, witness a group of young graduate students eager to take on the issues of globalization, perhaps filled with doubts about "Technology," that transcendentalized notion of a metaphysics that turns the whole world into "standing reserve." Perhaps they are also activists and ready to take to the barricades. What are their armaments? Laptops, e-mail, airplane tickets (obtained through some on-line service), and for entertainment in moments of boredom computer-run video games, or CDs playing the latest rock or techno music. These bright, young anti-globalists are themselves quite unlikely to pull the plugs on their own technologies.

The first trap likely reverberates with a second trap, which may be called the "nostalgia" trap. This is a form of criticism that, while worrying about changes made possible by technologies (only some of which will be followed), often contains a romanticized evaluation of life before the new technologies. Here is a cute example concerning a 1925–1926 set of worries about the onset of *radio broadcasting:*

> Broadcasting, it was claimed, would not only keep people away from the concert halls, it would stop them from reading books. It would encourage contentment with superficiality. "Instead of solitary thought," the headmaster of Rugby complained, "people would listen in to what was said to millions of people, which could not be the best things." Radio would make people passive. It would produce "all-alike girls." At the same time it would strengthen the forces making for healthy domesticity. "Charmed by the voice," husbands would stay at home in the evenings. Children would find a new source of daily satisfaction, shared no doubt by their mothers.[3]

In retrospect, these predictions concerning the dispersion of radio broadcasting seem far too vast, far too deterministic, and quite off the mark. But they do resemble the mindsets of those today who might be identified with "critical theory" and the notion that such technologies encourage "mass man" as opposed to the elite cultures of the capitalized "Kultur." Of course, the utopians concerning technologies can be just as wrong. Our local newspaper, *Newsday,* reissued on January 1, 2000, an issue of its predecessor newspaper from January 1, 1900, with glowing predictions that the then-new technology coming "on line," as we say, would transform New York City. That technology was *compressed air!* Extrapolating from the example of compressed air tubes that sped receipts from the department store sales counter to the central office upstairs for change, the paper's futurist foresaw a time when compressed air would be the power for underground transportation whose speed and efficiency would displace the horribly smelly steam locomotives that showered sparks from the "El" onto pedestrians below, and reduce the need for the many horse-drawn vehicles that clogged and be-shat the streets of the City. Both transportation and the U.S. Mail service, which would also be sped by compressed air tubes, would be technologically modernized by this compressed air development. And there were attempts made to bring this technology "on line" to the extent that at least regional compressed air tubes were put in place for the U.S. Mail service. These long-abandoned nineteenth-century mail tubes were recently rediscovered and are now being considered as possible conduits for a fiber optic system, with the usual legal battles concerning who may exploit this "right-of-way."

Now, for the nonce and for fun, take a quick look at a primary figure who combines the "functional contradiction" and "nostalgia" traps. I refer, of course, to Martin Heidegger, who is used as a dystopian prophet. Here is Heidegger on the hand and the typewriter:

> Human beings "act" through the hand; for the hand is, like the word, a distinguishing characteristic of humans. Only a being, such as the human, that "has" the word can and must "have hands." . . . [T]he human being does not "have" hands, but the hand contains the essence of the human being because the word, as the essential region of the hand, is the essential ground of being human. The word as something symbolically inscribed and as thus presented to vision is the written word, that is, script. As script, however, the word is handwriting.[4]

Now, enter technology: Note that writing technologies have changed, but remain technologies, since the pen, the quill, or the stylus for making cuneiform inscriptions, are also technologies. Rather, Heidegger hits upon a *modern* technology, the typewriter:

> The typewriter snatches script from the essential realm of the hand—and this means the hand is removed from the essential realm of the word. The word becomes something "typed." . . . Mechanized writing deprives the hand of dignity in the realm of the written word and degrades the word to a mere means for the traffic of communication. Besides, mechanized writing offers the advantage of covering up one's handwriting and therewith one's character. In mechanized writing all human beings look the same.[5]

Does Heidegger echo the previous "sameness" thought to be brought by the radio? But, consider two ironic phenomenological variations on this double trap. First, substitute a musical technology for the inscription technology: must Heidegger prefer the lyre or harp over the piano? For one plucks the lyre or harp with the "hand" and if the player is expressive, the delicacy of the tones surely do the same for music as the pen for writing. How terrible, then, to have "mechanized" the playing by moving into the evolution of keyboards, from harpsichords to claviers to pianos! The plucking is now "mechanical" and the fingers do not touch the strings. And there is historical evidence that precisely these objections were made with the introduction of the clavier. Yet, only one who is tone deaf would not be able to tell the vast distance between Horowitz or Ashkenazy and the beginner's woodenly played "mechanical" piece. The analogy with writing is no different—if the "letters" look the same (as they do with keys), the result of a typewritten piece by Nietzsche or a Mark Twain (two of the first who composed by and loved the typewriter amongst writers) and the local high school student is just as vast. What is expressed musically or written comes through the technology, but virtuosity shows. They look the "same" only if not actually read, hence, they look the same only to a superficial view. The Heideggerian "cheap shot" aimed at technologies, whether typewriters or constructed earth shots simply ignores the need to critically listen, read, or interpret results.

Then, in a second variation, which addresses the "functional contradiction" or pull-the-plug trap, let us resurrect our Heidegger and make him professor of a Stony Brook seminar—he demands that all research and term papers be *written by hand* for the reasons he gives. My suspicion is that our laptop generation of graduate students would quickly revolt and equally quickly regard our vaunted master as an "inscription Luddite," rather than the bold, Jeremiac prophet insightfully decrying technologies that they usually take him to be.

In the end, what the "functional contradiction" misplaces is the advantage which all materializations hold and which have been hinted at before. Plato, with his nostalgia about the verbal word and diatribe against writing—but who would not be remembered at all were it not for material manuscripts—may have been the first in our tradition to exemplify this functional contradiction. The same would apply to Heidegger, had he to have written each copy of each book of his by hand (and had he done so, might we have better known his character?). How limited his fame—again through the functional contradiction—were this to have been a requirement of dissemination.

I actually agree with Heidegger that one "acts" through the hand better, through one's being a body. But the technologies that can materialize expressiveness are many and varied. The sense of body-machine embodiment experienced through a word processor does not, in my estimation, "degrade" the hand. It merely embodies it differently than through the dip pen. These variations resist what I take as the extremes of both utopian and dystopian interpretations of what technologies imply.

These two traps are not the only ones that lie along the path to globalization, but they are at least symptomatic of forms of resistance that I believe to be spurious, but which appear over and over again. On the other hand, neither is some linear, deterministic outcome to be expected regarding the technologies that make globalization possible. Compressed air never attained its promise, and the fact that its material infrastructure is about to be "translated," as Latour would say, into pathways for fibre optics shows that technologies, too, change along the way. There is always a certain unpredictability that accompanies all technological trajectories. And, just as the Heideggers get caught in nostalgia and contradiction traps, utopians and determinists get caught in "designer fallacies."

Heidegger's famous hammer indeed "withdraws" in use, becomes embodied by the user, and through it, in action, accomplishes some project in the world. But we don't really know from his text if Heidegger's hammer is cobbling shoes, shingling a house, or placing a nail in a wall to hang a picture. And, in spite of design differences between shoe cobbler's hammers and carpenter's claw hammers, each of these could also be paperweights, murder weapons, or objets d' arte. The "same" artifact can belong to a multiplicity of contexts with all their assignments and no design can "determine" use. If this degree of polyvariance adheres to a simple tool, such as a hammer, the same and more adheres to a complex technology such as our aforementioned Internet.

Originally designed to facilitate communications—often encrypted—between elite military and university researchers in the heat of the cold war, today the same Internet allows pedophiles to cruise chat rooms, dot com scammers to con victims, advertisers to spam wider audiences, hackers to challenge and distribute thousands of varieties of viruses, trojans, and worms, and in an analogue to minitel such as Andy Feenberg has pointed out, rather than becoming a commercial enterprise, it has turned out to foster romantic connections, and been used for gossip, support groups, ad infinitum. And this same Internet allowed me to find the Heidegger quotation about the hand, which I had forgotten to bring along from Long Island to Vermont in my annual summer move. If the Internet was designed, it is design out of control. Put more strongly, *it is impossible to design limited use into any technology.* This suggests that the same applies to the phenomenon of globalization. And with this we return to the beginning implication of the title: we are already being globalized—over that we have no choice—but globalization is much squishier than claimed by, or intended by, either its detractors or its promoters. Its technologies hold forth both promises and unsuspected results and outcomes. Does that mean we can "choose" the kind of globalization we want?

Not quite. Both pulling the plug and designing the result belong to the same deterministic and rationalistic error. Just as the death of Cartesianism eliminates the ideal rational observer or Cartesian god, it also eliminates the god trick *controller* of destinies. What we need is a model of analysis that recognizes globalization as a change of texture in a lifeworld. Or, alternatively, to see these changes along the lines of a Latourean actor network which encompasses both the humans and non-humans, as a process in which we take up *responsible non-innocence* such as Donna Haraway suggests. Each of these approaches allows the problems of globalization to remain recognizable: what is the relationship between "free market" or capitalistic economic interchange and representative democracies? Does globalization imply only third world to first world flows? Or is the opposite flow also possible? Does globalization imply a "standarization" of cultures? Or a pluralization of cultures? Does globalization lead to a system of postnationalistic justice? Or do political cultures remain autonomous? The answers to all these questions must be, I believe, ambiguous.

I suspect the dynamics of globalization are amplified echoes of some earlier human developments. Earlier historians and anthropologists usually describe the movements from earlier human communities into a "civilization," as one which included domestication of plants and animals, development of writing, cities, laws, trading practices, and the like. In this much earlier transformation of human communities, there can be seen a forecast concerning "standarization," for example. Take domestication: domesticated food sources clearly displayed a reductive movement from the vast number of grains harvested by hunter-gatherers (Australian Aborigines harvested more than thirty grains and made breads from each) to the few that were selected and bred, usually fewer than a half-dozen grains. And, even here, we recognize dominant grains such as wheat cultures (Middle East and Europe), rice cultures (Asia), corn cultures (the Americas), etc. I cannot here belabor both the losses and gains which this early "standarization" made, but in today's world there can be and are countermoves that can be undertaken as well. American Land Grant agricultural universities are experimenting with hypertrophizing an ever-widening set of wild grains, partly to overcome the disease weaknesses prevalent with monocultural grains, and partly to explore the variety of tastes and nutritional properties possible. The very same reduction of animals domesticated—usually a half-dozen or so—compared to the variety of nondomesticated ones, occurred in antiquity and remained to contemporary times. (I cannot resist at least imagining a Mesopotamian Heidegger decrying the irrigation of the river valleys of five millenia ago!)

In the opening examples the growth of awareness of the perspectival notion of earth-as-planet was cited. Earth is now visually presented in gestalt fashion in space shots, shots from the moon, etc. Also, the globalization of alternate face/virtual communication and meetings made possible by contemporary communication and transportation technologies was highlighted. In both cases the focus was upon the materiality of the technologies that enable these aspects of globalization. But what occurs in a focus is not all that occurs. Any technology, deployed, carries with it "side effects," many of which are unpredictable, and which are often both positive and negative. Here are some examples of precisely these material side effects in the context of globalization history.

To continue with my penchant for irony, let me begin with one forefront set of debates that relate to globalization in its popular, media sense, that is, a debate that catches up internationalization, large corporate multinationals, and futuristic technological developments—I refer here to "GM" biotechnologies. Genetically modified organisms result from the successful technological manipulation of microstructures of plants and animals made possible by the biotechnological techniques developed only in the last few decades. With such new techniques, GM products immediately fall under the aura of future fears and hopes already noted. "Frankenstein" technologies run amok, intruding into god-space-variants upon the dystopic descriptions of radio already noted, are matched by promises of overcoming all hunger, the perfectly cloned animal, perhaps the perfect child, and at the least, a pushing back of genetically related diseases such as Alzheimers, Parkinsons, and juvenile diabetes. These are bound to and already have come forth in patterns philosophers of technology well recognize.

Admittedly, no one knows what effects genetically modified organisms will have—no more than did our ancestors know what the domestication of plants and animals would have upon their futures. And in both cases those who fear and those who utopianly hope will again fall into line with the old arguments we have already heard. I will return later to this problem of ignorance and unpredicability, but for the moment the concentration is upon an interesting result with respect to globalization disputes, which poses an interesting symptom for our dilemma about what globalization we want.

That we should have a serious discussion and consideration concerning genetic modification goes without saying. But why, how, and in which way we should have this critical consideration is also part of how we can incline toward desirable rather than undesirable globalization as well.

The GM fight is already on. From Pacific Northwest "eco-terrorism," which burns crops and attacks trees, to the anarchism that burns SUVs and applies arson to new developments (some on Long Island), the attempt to stop certain directions of change makes for the usual juicy media news. But all of this leaves aside an extant, massive, expensive, and environment-threatening set of effects that receive at best background notice and which in terms of effects are already massively telling.

Edward Tennant, in a remarkable book, *Why Things Bite Back*, does a history of unintended side effects to technologies, including those that relate to agriculture and, potentially, biotechnology. Humans, probably as long as they have moved around upon the face of the globe, have often taken along with them a biological entourage—migrating peoples have taken their pack animals (horses, oxen, camels) and pets (dogs, cats, birds), and while the histories on these are dim, they have had massive effects upon the environments into which these beasts and plants have been introduced. Example: European colonists introduced (technically, reintroduced, since there were equine species in North America millions of years earlier) the horse to the Americas. The horse, subsequently domesticated by native North Americans, dramatically transformed what had been once been forest–edge dwelling peoples, who occasionally got a buffalo, into the Plains Indian tribes, which became migratory and followed the buffalo.

And, this penchant for taking the familiar with us stimulated our European ancestors to introduce the English sparrow, the European starling, domestic cats, and the like, along with Bulgarian wheat, milo soybeans, and later kudzu and the like into the Americas with results we all know. "Sparrows rule!" Starlings have mostly pushed out the eastern bluebirds, kudzu fills the southern U.S.. And, in Australia, the domestic cat gone feral has spread into the hinterlands where no predator like it existed before and is systematically driving ground-nesting birds to extinction. All these examples are pre-twentieth-century globalization effects.

The very way in which these lists are compiled, however, can also bias the effect. I have listed above a set of pests, monocululture plants, and predators gone wild. I could have simply listed what we would likely consider to be so familiar and taken for granted, that we forget that the flora and fauna were likewise transported and introduced. What would we, or Europe, be without potatoes, tomatoes, beans, pumpkin and squash, and other foodstuffs that came from the Americas? Or, as a recent book has shown me, what is more "American" than *apple pie,* yet apples came from Kazakhstan, were crossbred and introduced to North

America by the English—along with the domestic honey bees needed to pollinate the trees! In short, any new introductions, technological or biological, have mixed and ambiguous results.

More sinister are the effects of *unintended* biological infestations, which, I remind you, include HIV as an import, between species (monkey to human) and continents. Or, take the now two-year-old phenomenon of the West Nile Virus, becoming endemic and spreading, largely through crows, in the Northeastern United States, another probably airplane-borne infector from the Middle East. From the far Pacific, poisonous brown tree snakes, presumed hidden in the landing gear of aircraft originating from their indigenous areas, have made their way to the Hawaiian Islands and are, like the cats in Australia, decimating bird populations. The "globalization" of biological infectors, both plant and animal, such as zebra mussels clogging the drains of the northern streams and lakes, which followed the earlier lamprey eels, supplemented by Asian eels now spreading through the South, are all part of an exceedingly damaging and expensive set of new arrivals (most unintended hitchhikers on contemporary transportation devices). Later on, the invaders may be restrained—the gypsy moth infestations are now recognized to occur in cycles, and introduced viruses that attach the larvae are usually successful. With all this, where are our demonstrators? Where are the anarchists now that we need them? Admittedly, in the current context, Monsanto may appear as a more easily attackable enemy than "nature" displaced, and given our pre-prepared images of the movie mad scientist, the genetic manipulator seems more easily locatable. Yet, in all this, we remain hopelessly anthropomorphic in our targeting practices and overly simplistic in terms of our pre-globalized lifeworld.

We must also allow the utopians to have their say against this set of eco-disasters just outlined. Who today fears the scourges of polio, cholera, diptheria, syphillis, or tuberculosis? For even where not eliminated, the battle so far remains in our modern favor. Are there not also more democracies today than five decades ago? And the glimmerings of an international, humanistic justice system may be appearing in the internationalization of justice in which Pinochet and Milosevic, once heads of state, get caught for antihumanitarian crimes, dragged against their outdated nationalist wills into an international forum. Even life expectancies and birth rates are respectively up and down, contrary to the neo-Malthusian predictions of only three decades ago. These, too, are side effects of globalization, paralleling those of the invasions noted previously.

Where does this leave matters? Pretty much at precisely the state of the opening problem of whole earth measurements. The phenomenon of globalization is obviously too complex, too ambiguous, and too much part of our lifeworld already to have one clear and distinct shape. Nor is it possible to deal with a complex global phenomenon on the basis of past "rational control" epistemologies. So, having dimly perceived the phenomenon, having begun to recognize that we are non-innocently part of it, what do we do about it? At this point I have to be daring, but in a rather pragmatic and fallibilist way. I shall suggest that in parallel to technologies, which also turn out to be complex, unpredictable and multivariant, and which—if one wants to change them—one has to conceive of as something like a *culture,* globalization must be similarly conceived. And, before taking up pragmatic, fallibilistic suggestions, just think for a moment how hard it is to change a culture. What follows are some heuristic suggestions for living in and non-innocently trying to incline facets of globalization toward the shape we want.

First, where should we position ourselves? My answer is that those who wish to nudge the direction globalization is going to take must position themselves at "Research and Development" junctures. This is a lesson I have learned with respect to the problems of ethics related to technologies. The usual—and even the societally expected—role for the ethician is usually, contrary to this suggestion, to be positioned at the end of a development process, after the technology is already in place. For example, in medical ethics, the early role of ethicians was that after a new technology was developed, then the ethician stepped in and helped with understanding and distributing results. This, however, is a sort of "band-aid" ethics, to put the bandage on after the wound is discovered. Rather, "R & D" positioned thinkers must think critically about processes in the making. Again, drawing from current debates, stem cell research is particularly interesting both as a problem and as an indicator of precisely a rising awareness about the importance of R & D positioning. I take it as a healthy indicator that both scientists and politicians are beginning to recognize the complexity and the promise of this line of research. The previous coalition of simplistic approaches—thou shalt not use any embryo stem cells, but maybe one can use bone and marrow adult stem cells because the former encourages abortion, whereas the latter may not—is rapidly breaking down. I take it this is a small sign that a tiny bit more understanding of possibilities and implications is becoming known. It is also a sign that the need to consider developmental problems prior to implementation is beginning to be considered. This argument at least occurs today at the beginnings of a research development, rather than at the end of one such as the introduction of kidney dialysis machines, which produced the first ethics committees in many university hospitals.

Second, to be positioned in a R & D position calls for much more awareness on the part of the critics of the intricacies of the process than mere layman's knowledge. The R & D positioned person must be at least an informed amateur regarding the science involved. Unfortunately, the North American context for science and technology assessment is often structured precisely wrong for this *critical hermeneutic* role to be prominent. The agonistic juxtapositioning of scientist/lay public which too often characterizes our situation simply helps feed the extremism of positions such as those taken by eco-terrorists. Donna Haraway, Andrew Pickering, and I have all pointed out that the Northern European approach to developmental assessment is much more congenial to producing the possibilities for critical hermeneutic stances. (I refer to Dutch, German, and Scandinavian approaches primarily.) For example, in Holland the government has decided it is extremely urgent that there be a national consideration and discussion of GM foods. It has, following previous practices, identified and mandated that the various interest groups (consumers unions, agriculturalists, political groups, educators, etc.) be identified and in conjunction with the research community undertake mutually educational and political assessment discussions. As it turns out, a hermeneutically trained philosopher, Bart Gremmen, has been named the coordinator of this process.

(From my own indirect involvement with a previous problem of genetic manipulation, involving "Herman the Bull," who was engineered to produce a strain of anti–lactose allergenic genes, the results can be surprising. It was a democratic decision *not* to let Herman have progeny, just as earlier in a similar decision in Germany, there was a determination that nuclear power would not be utilized for electricity generation. In each of these cases, some of the reasons might be culturally interesting to us: In Herman's case, one reason was that "cowness," a very important Dutch value, was not being properly respected; and in the nuclear case, one reason was that there was not a level playing field to make such a decision because not proportionately enough research money had gone into non-nuclear possibilities, and until such occurred there was not sufficient reason to support nuclear options.)

Third, wherever positioned, it is important that the methods, particularly the ethical and political methods used to make the inclinations themselves, be of the highest standards. The question always remains the longer term one—what processes do we want to have used both before and after any direction is taken? It is here that I part company with any form of terrorism or violence-inducing politics, including eco-terrorism. The Black Bloc of the recent G-8 sessions use methods that are no different than than those used by the Pinochets and Milosevics, or, to put it starkely, eco-Milosevicism is equivalent to Milosevic-Milosevicism. This confrontational and destructive method has already led to loss of life

and will lead to more, whether intended or not, whether caused by
terrorists or the stopping of terrorism.

(I am reminded of my own earlier 1968 experiences in France in
the midst of the "Days of May" strikes. I had been highly sympathetic
to the student-worker movement, until the street conflicts got to the
point that the rebels began to chainsaw the ancient sycamores along the
boulevards to slow police movements, an action that made me begin to
lose sympathy then, since the destruction of ancient trees seemed excessive, just as in a recent situation, one can detect little sympathy for the
Taliban destruction of ancient Buddhas in Afghanistan.)

Fourth, and finally, in all this I suggest that none of us are wise
enough to predict outcomes and all of us are going to be surprised by
some outcomes, thus calling for a lot of humility regarding even our
highest values, rather than arrogance. Here I find myself sympathetic to
a kind of neo-Enlightenment set of values. For example, one of the
inheritances from Enlightenment values is religious tolerance. Note how
long it took for this to be written into modern secular states—after
centuries of wars, after breakaway colonies such as our own, and most of
all, after the changes within religions themselves, religious tolerance has
become a rule of law. Today we are going through a similar process with
respect to ethnicity and, at least to my mind and experience, the most
successful locations where cultural tolerance takes place are precisely those
countries where religious tolerance is most strongly fostered. But one
must not forget, that for religious and cultural tolerance to work at all,
the religions and cultures themselves have also to change, to drop at least
the practices of absolutism and fundamentalism if not the subjective
beliefs about absolutism as such.

There may be a hint in this earlier experience for what will happen
to our senses of community and citizenship here, too. Many years ago
a close friend, Harvey Cox, wrote a book called *The Secular City*. In it
he noted that in terms of friendship, one is much more likely to be
friends with those who have similar interests and tastes in a widely distributed, rather than geographically close setting. That idea seemed, not
only true but liberating. One does not usually have all, or maybe even
many, of one's friends in the same apartment house or village in the
contemporary world. Rather, one may have widely dispersed friends,
with whom one communicates and visits in another variation of virtual/
face-to-face occasions. This is much more a quasi-urban model than a
"global village" model; it is a cosmopolitan model.

None of these suggestions can or should be taken as universal,
normative rules, but as pragmatic, flexible, and open-to-change guidelines. And all of these suggestions must be couched within a critical and

somewhat skeptical attitude. Hype, wild linear extrapolations, both dystopian and utopian projections, usually turn out to be as relevant as the examples of compressed air and radio broadcasting I have used.

And, as for the question of whole earth measurements, here is part of the answer. The images used to display such measurements, a depicted global display with false color projections showing imaged data, are constructs of millions of measurements taken from satellite passes, sea surface measurements, and laser measurements, producing a model that may be one of the most expensive "pictures" ever produced, costing literally millions of dollars; and, as one can see through informed vision, ocean surface waters rose 1.6 cm. between 1993 and 1994, according to *Science*.[6]

Then, with regard to the human scale, we are all familiar with published photos of people from every culture—I recall one showing a Masai warrior, carrying a spear and wearing a loincloth, holding and talking into a cell phone—taking up the technologies through which and with which we concretely experience globalization.

Notes

1. Don Ihde, *Expanding Hermeneutics: Visualism in Science* (Illinois: Northwestern University Press, 1998), 51–52.

2. Ibid., 57–58.

3. Asa Briggs, *A History of Broacasting in the United Kingdom* (Oxford: Oxford University Press, 1961).

4. Martin Heidegger, "On the Hand," cited in Michael Heim, *Electric Language* (New Haven: Yale University Press, 1987).

5. Ibid.

6. *Science*, 8 September 1995.

6

Looking Backward, Looking Forward

Reflections on the Twentieth Century

Andrew Feenberg

In the year 1888, Edward Bellamy published a prophetic science fiction novel entitled *Looking Backward: 2000–1887*. Bellamy's hero is a wealthy Bostonian who suffers from insomnia. He sleeps hypnotized in an underground chamber where he survives the fire that destroys his house. Undiscovered amidst the ruins, he dozes on in suspended animation for more than a century, awakening finally in the year 2000 in a Boston transformed into a socialist utopia. Most of the book is taken up with his puzzled questions about his new surroundings and his hosts' lucid explanations of the workings of an ideal society.

Bellamy's book is now forgotten except by specialists but it quickly became one of the best-sellers of all times, read by millions of Americans from the closing years of the nineteenth century until World War II. It articulated the hope in a rational society for several generations of readers.

In 1932, less than fifty years after Bellamy's famous book appeared, Aldous Huxley wrote *Brave New World*, a kind of refutation of *Looking Backward*. In the exergue to Huxley's book the Russian philosopher

Berdiaeff regrets that "utopias appear to be far more realizable than used to be believed." Berdiaeff goes on: "[A] new century is beginning, a century in which intellectuals and the cultivated classes will dream of the means of avoiding utopias and returning to a less 'perfect' and freer non-utopian society." Unlike *Looking Backward, Brave New World* is still widely read. It is the model for many later "dystopias," fictions of totally rationalized societies in which, as Marshall McLuhan once put it, we humans become the "sex organs of the machine world."[1]

We can now literally "look backward" at the twentieth century and as we do so, the contrast between Bellamy's utopia and Huxley's dystopia is a useful one to stimulate reflection on what went wrong. And, clearly, something very important did go wrong to confound the reasonable hopes of men and women of the late nineteenth century. While they expected social progress to proceed in parallel with technical progress, in reality every advance has been accompanied by catastrophes that call into question the very survival of the human race.

What happened to dash those hopes? Of course we are well aware of the big events of the century such as the two world wars, the concentration camps, the perversion of socialism in Russia, and more recently, the threats from genocidal hatreds, environmental pollution, and nuclear war that we carry with us from the last century into this one. But underlying these frightful events and prospects, there must be some deeper failure that blocked the bright path to utopia so neatly traced by Bellamy.

Could a spiritual flaw in human nature or in modernity be responsible for the triumph of greed and violence in the twentieth century? No doubt human nature and modernity are flawed, but this is old news. Bellamy and his contemporaries knew all about greed and violence, the insatiable appetites, the pride and hatred lurking in the hearts of men. They understood the battle between Eros and Thanatos as much or as little as we do. What has changed is not our evaluation of human nature or modernity but the technical environment that has disrupted the delicate balance between the instincts that still left Bellamy's contemporaries room for hope, indeed for confident predictions of a better future.

We can begin to understand this technical shift by considering what is missing from Bellamy's description of society in the year 2000. His world is completely industrialized, with machines doing all the hardest work; improved technology and economies of scale have raised productivity to the point where there is enough of everything. Workers are drafted into an "industrial army" where a combination of expert command and equal pay responds to the claims of technical necessity and morality. Although this is clearly an authoritarian conception, it is important to keep in mind that obedience is ethically motivated by the economic equivalent of military patriotism, rather than imposed through management techniques. Workers can freely choose their jobs after a brief period of manual labor at the end of regular schooling. Labor supply is matched voluntarily to demand by offering shorter workdays for less desirable jobs.[2] Workers retire at forty-five and devote themselves to self-cultivation and to the duties of full citizenship, which begin at retirement.

Bellamy's utopia is essentially collectivist, but, paradoxically, members of the society are depicted as highly differentiated individuals, each developing his or her own ideas, tastes, and talents in the generous allotment of leisure time made possible by technological advance. Individuality flourishes around the free choice of hobbies, newspapers, music and art, religion, what we would call "continuing education," and democratic participation in government. Invention, too, appears as an expression of individuality and a source of social dynamism.

None of these activities are organized by the industrial army because there is no scientific-technical basis for any of them, hence no technology requiring expert administration, and no objective criteria of right and wrong, better or worse. The economies of scale that make industrial technology so productive in Bellamy's account have no place in these activities, which depend on individual creativity.

Those who wish to act in the public sphere through journalism, religious propaganda, artistic production, or invention therefore withdraw from the industrial army as they accumulate sufficient "subscribers" to their services to justify the payment by the state of a regular worker's wage. The state provides these cultural creators with basic resources such as newsprint without regard for the content of their activities.

How different this imaginary socialism is from the real thing as it was established in the Soviet Union only a generation after Bellamy published his book! His society is bipolar, half organized by scientific-technical reason and half devoted to Bildung, the reflectively rational pursuit of freedom and individuality. But this bipolarity is precisely what did not happen in the twentieth century under either socialism or capitalism. Instead, total rationalization transformed the individuals into objects of technical control in every domain, and especially in everything touching on lifestyle and politics.

It is interesting to see how close Bellamy came to anticipating mass society. At a time when phone hookups still numbered in the thousands, he imagined a telephone-based broadcasting network which, he predicted, would disseminate preaching and musical performances. Each house would have a listening room and programs would be announced in a regular printed guide. Bellamy even understood that musical performance in the home would decline as broadcasts by professionals replaced it. So far his extrapolations are remarkably prescient, but nowhere did Bellamy anticipate the emergence of gigantic audiences subjected to commercial and political propaganda. Nor did he suspect that the small publications of his day, individual artistic production, and personal preaching would be so marginalized in the future that they would be unable to sustain the individuating process that was for him the ultimate goal of social life.

Higher culture, both religious and secular, had a moderating and civilizing effect in his century, so enlarging the space of its influence through a more generous provision of education and leisure promised social advance. This, and not wealth as such, was the reason for Bellamy's optimism. In his vision standardization and control were confined to the struggle with nature. What Norbert Wiener called "the human use of human beings" was apparently unthinkable. But the creation of the mass audiences of the twentieth century continued the industrial pattern of efficiency through economies of scale in the application of technology. The twentieth century saw the displacement of higher culture in public consciousness by a mass culture dedicated to unrestrained acquisitiveness and violent political passions.

Brave New World, on the other hand, was written a decade after the first commercial radio broadcasts, which adumbrated a future of mass media manipulation. Huxley's vision was simply extrapolated from the rise of modern advertising and popular dictatorships. In *Brave New World,* the radical overextension of rationalization makes human beings into willing servants of a mechanical order. The Marxist hope, which Bellamy shares, for human mastery of technology no longer makes sense once human beings have themselves become mere cogs in the machine. This same view underlies much twentieth-century thought, for example, pessimistic social theories such as Max Weber's and the various deterministic philosophies of technology influenced by Martin Heidegger.

Heidegger's concept of enframing describes a state of affairs in which everything without exception has become an object of technique. Things are now defined by their place in a methodically planned and controlled action system. All being is raw materials in technical processes; nothing stands before being as the place of awareness. Complete meaninglessness threatens where the unique status of human "Dasein," as the being through which the world is revealed, is so completely denied.

Heidegger might be thought of as the philosopher of *Brave New World,* except that he would deny that what we have before us today is a "world" in the full sense of the term. Rather, we are surrounded by an "objectless" heap of fungible stuff that includes us. This deep pessimism and a certain moral insensitivity are reflected in his shocking statement to the effect that "[a]griculture is now the mechanized food industry, in essence the same as the manufacturing of corpses in gas chambers and extermination camps, the same as the blockade and starvation of nations, the same as the production of hydrogen bombs."[3]

After Heidegger, a number of other philosophers developed similarly pessimistic views of modern society. The Frankfurt School philosopher Herbert Marcuse was a student of Heidegger's and his critique of "one-dimensional society" resembles his teacher's thought in a Marxist guise. Heidegger distinguished between craft labor, which brings out the "truth" of its materials, and modern technology, which incorporates its objects into its mechanism under the domination of a will and a plan. In Marcuse this Heideggerian approach continues essentially unchanged as the distinction between the intrinsic potentialities of things, which might be brought out by an appropriate art or technique, and the extrinsic values to which they are subordinated as raw materials in modern production. And like Heidegger, Marcuse deplores the extension of the latter approach to human beings themselves.

But unlike Heidegger, Marcuse holds out the possibility in principle, if not much hope, of creating a new technology that respects the potentialities of human beings and nature. This "technology of liberation" would be a "product of a scientific imagination free to project and design the forms of a human universe without exploitation and toil."[4] This is still a worthy goal, although perhaps it should be described as a receding horizon: today we seem to be as far from achieving it as when Bellamy wrote.

These are what I call dystopian philosophies of technology. They had surprising influence in the 1960s and 1970s despite their notorious difficulty. Dystopian themes showed up not only in politics but in films and other popular media, discrediting liberalism and gradually infiltrating conservatism as distrust of "big government." Contemporary politics is still strongly influenced by vulgarized versions of the dystopian sensibility. These changes were accompanied by a dramatic shift in attitude toward technology. By the end of the 1960s technophobia had largely replaced enthusiasm for nuclear energy and the space program. No doubt the arrogance of the technocracy and the absurdity of the war in Vietnam played a major role in this change.

As dystopian consciousness spread, it was transformed. No longer a theoretical critique of modernity, it inspired a populist movement that rejected its own cultural elitism. The question of technology was now a political question. The New Left reformulated socialist ideology in a tense compromise between traditional Marxism and the protest against dystopia. In so doing, it opened a space for the new technical politics of recent decades, which engages in concrete struggles in domains such as computers, medicine, and the environment.

The French May Events was by far the most powerful New Left movement, the only one with massive working-class support. In the spring of 1968, France was paralyzed by a general strike inspired by a student protest. Some ten million workers walked off the job and closed down the entire economy and most of the government, threatening the capitalist system. The May Events was an antitechnocratic movement, as hostile to Soviet-style socialism as to advanced capitalism. The students and militant workers proposed self-management as an alternative to planning and markets. Their position was summed up in a widely circulated leaflet: "Progress will be what we want it to be."[5]

The movements of the '60s undermined technological determinism, both in theory and practice. But they continued to employ a dystopian rhetoric in response to the technocratic threat. However, as the twentieth century came to a close dystopianism lost much of its authority. Journalists and science fiction writers devised new utopias inhabited by bioengineered superhumans networked in a universal mind or downloaded to more durable hardware than the human body. Technology plays a central role for Bellamy and Huxley, but the advances they describe are symbols of hopeful or disastrous social trends rather than specific technological forecasts. These contemporary utopias are presented as breathless frontline reports on the latest R and D. Determinism returns as social consequences are deduced from future technology. Serious thinkers perplexed by this upsurge of horrific speculation once again raise flimsy ethical barriers to "progress." Dystopian humanism struggles to salvage spirit from the Satanic mills of advancing technology. But the whole contest begins to seem routine and not very credible.

Meanwhile, new trends have emerged among researchers who eschew speculation and study technology as a social phenomenon. These researchers view the dystopian critique of modernity as nostalgic longing for a past that is forever lost and that was not so great in any case. According to this view, we belong wholly and completely to the technological network and do not represent, nor should we await, a suppressed alternative in which "man" or "Dasein" would achieve recognition independent of his tools.

Non-modern or posthumanist thinkers such as Bruno Latour and Donna Haraway have put forward this revisionary approach with singular energy in books and essays with titles such as *We Have Never Been Modern*,[6] and "The Cyborg Manifesto." The very tone of these titles announces an agenda for the new millennium. According to the authors, we have passed through the experience of dystopia and come out on the other side. Our involvement with technology is now the unsurpassable horizon of our being. No longer opposed to technology, we join together with it in a more or less undifferentiated "cyborg" self.[7] It is time to cease rearguard resistance and, embracing technology once and for all, give its further development a benign direction.

The Internet supplies the essential social background to the wide interest in this posthumanist view. Of course the authors did not have to go online to develop their ideas, but the popular credibility of their innovative vision depends on the emergence of computer networking and the new function of subjectivity it institutes. Without the widespread experience of computer interaction, it is unlikely that their influence would have spread beyond a narrow circle of researchers in science studies. However, given that experience, they articulate a fundamental shift in the relation of human beings to machines, from antagonism to collaboration.

What is it about networking that assuages dystopian consciousness? The fear of dystopia arises from the experience of large-scale social organization which, under modern conditions, possesses an alienating appearance of rationality. The loss of individuality is exemplified in the relation of mass audiences to the new media of the twentieth century until computer networking breaks the pattern. Instead of the passivity associated with participation in a broadcast audience, the on-line subject is constantly solicited to "interact" either by making choices or responding to communications. This interactive relationship to the medium, and through it to other users, appears nonhierarchical and liberating. Like the automobile, that fetish of modernity, the Internet opens rather than closes vistas. But unlike the automobile, the Internet does not merely transport individuals from one location to another; rather, it constitutes a "virtual" world in which the logic of action is participative and individual initiative supported rather than suppressed by technology. This explains the proliferation on the Internet of expressions with the pronoun "my," as in "My Yahoo," "My MP3," and so on.

It is noteworthy that this evolution of the network owes more to users than to its original designers, who intended only to streamline the distribution of information. Refuting technological determinism in practice, users "interacted" with the network to enhance its potential for interaction. This was the real "revolution" of the "Information Age" which transformed the Internet into a medium for personal communication.[8] As such it is a switched system like the telephone in which the corporate giants who manage the communication have little or no control over what is communicated. Such systems, called "common carriers," extend freedom of assembly and so are inherently liberating.

What is more, because computer networking supports group communication the Internet can host a wide variety of social activities, from work to education to exchanges about hobbies and the pursuit of dating partners. These social activities on the Internet take place in virtual worlds built with words. The "written world" of the Internet is indeed a place where humans and machines appear to be reconciled.[9]

At this point, a note of caution is in order. The enthusiastic discourse of the information highway has become predictable and tedious. It awakens instant and to some extent justified skepticism. It is unlikely that the twenty-first century will realize the dream of a perfectly transparent, libertarian society in which everyone can work from their home, publish their own book, choose multiple identities and genders, find a life partner on-line, buy personalized goods at an electronic mall, and complete their college education in their spare time on their personal computer. It is reasonable to be suspicious of this vision. The dystopian critic finds here merely a more refined and disguised incorporation of the individual into the machine.

But both utopian and dystopian visions are exaggerated. The Internet will certainly have an impact on society, but it is ludicrous to compare it with the Industrial Revolution, which pulled nearly everyone off the farm and landed them in a radically different urban environment. My "migration" to cyberspace over the last twenty years can hardly be compared with my ancestors' migration from Central European villages to New York. Worrisome though it may be, the "digital divide" is far more easily bridged than the divide between city and country in a society without telephones, televisions, and automobiles. Unless something far more innovative than the Internet comes along, the twenty-first century will be continuous with our world, not a radical and disruptive break. Its real significance lies not in the inauguration of a new era, but in the smaller social and technological changes it makes possible at the current level of advance.[10]

The political question is not whether the Internet will liberate or enslave us, as though a technology had that power, but rather the subtle change in the conditions of public organization and activity it promotes. This change had already begun before the rise of the new medium, but intermittently and laboriously.

The Internet promises to enhance the ability of the population to intervene in the technical decisions so vital in a society like ours. So long as citizenship is defined by traditional geographical districts, its influence on technical life is severely limited. What can a local community do about a technology that crosses all boundaries, for example, a new medicine or a new method for producing food? The "public," which ought in principle to be able to comment on such changes and influence them democratically, is not locally defined. It is fragmented into subgroups that follow the lines of specific technical mediations. Thus, the "citizen" is more and more a factory or clerical worker, a student or teacher, a victim of a disease, a consumer of industrial food products, and so on.

Eventually, these technological citizens may find their needs and interests represented by traditional geographically based politics, but not before many struggles and protests have prepared the issue on the terrain of technically mediated subgroups. These struggles and protests first define the new issues and bring them to the attention of a larger audience. But the task is unusually difficult. Technical publics are fragmented and the problems they confront unfamiliar. The creation of a technical public sphere is thus arduous and uncertain.

It is interesting to note that John Dewey had a better grasp of this situation in the 1920s than many contemporary political philosophers. He worried that traditional local community was losing its integrity in a mobile modern society. New forms of technically mediated community were needed to replace or supplement localism, but these were not easy to create. Dewey described the dilemma as follows:

> Indirect, extensive, enduring and serious consequences of conjoint and interacting behavior call a public into existence having a common interest in controlling these consequences. But the machine age has so enormously expanded, multiplied, intensified and complicated the scope of the indirect consequences, has formed such immense and consolidated unions in action, on an impersonal rather than a community basis, that the resultant public cannot identify and distinguish itself.[11]

What the Internet has done is to make it much easier for these publics to engage around the technical mediations by which they are shaped. To be sure, the Internet itself is not essential to this evolution and the mere existence of the technology does not guarantee any particular usage. Before computer networking took off, technical publics emerged around other issues such as nuclear power, environmental pollution, and the AIDS crisis. In these cases too, participants in one or another technical network linked up to achieve political and technical changes. But the very exceptional nature of these occasions, and the extraordinary difficulty of putting together the long chains of activists scattered over huge territories suggests the potential significance of the Internet.

New communication technology enables far more precise, detailed, and convenient coordination and control of business activities at a distance. Administrative and managerial elites can use the network to build disseminated organizations that evade local, indeed even national, restrictions and power structures. Unless citizens can also coordinate and internationalize their movements, they risk becoming totally irrelevant. Imagine the consequences if corporations and governments alone had access to the network and ordinary people remained just as provincial and cut off from each other as in the past. The disproportionate power of large-scale organizations would be irreversibly enhanced. Today, democracy depends to a significant extent on public mastery of the network.

The first persuasive evidence of the unusual democratic potential of the Internet came from a surprising source: the Zapatista movement in impoverished Chiapas brought its struggle to world attention on the Internet, blocking a violent repression that might have suppressed resistance there for another generation. The anti-globalization demonstrations in Seattle, Washington, Prague, and Genoa have shown the power of the Internet in a more modern context. International protest now corresponds to international finance and trade. We can expect ever more of the same in fields such as medicine, education, and the environment. Let me reiterate: this is not a claim that the Internet will liberate us, but rather that it will make it considerably easier to address the problem that worried Dewey, the inability of geographically dispersed technical publics to articulate their concerns.

In conclusion, I would like to discuss briefly some philosophical approaches to understanding these new forms of struggle. As we have seen, the Internet supports a vision of harmonious coexistence between humans and their machines. But these political applications pinpoint something else that was well understood by dystopian thinkers. They argued that technology is a source of power over human beings and not merely an instrument for the satisfaction of human needs. Because that power is essentially impersonal, governed by technically rational procedures rather than whims or even interests in the usual sense of the term, it appears to lie beyond good and evil. This is its dystopian aspect.

What Marcuse called one-dimensionality results from the difficulty of criticizing the "system" in terms of traditional concepts of justice, freedom, equality, and so on. But we have seen that the exercise of technical power evokes resistances immanent to one-dimensional society. Technological advance unleashes social tensions whenever it slights human and natural needs. The narrowness of its social and economic base ensures that such slights occur often. After all, the system is not a self-contained expression of pure technical rationality but emerged from two centuries of deskilling and abuse of the environment under the pressures of capitalist competition and bureaucratic socialism. Vocal technical publics arise around the tensions caused by these limitations. Demands for change reflect aspects of human and natural being denied by the dominant structure of the technical system, what I have called its "technical code."[12] Here dystopia is overcome in a democratizing movement the full extent of which we cannot yet measure.

A modified version of the system/lifeworld distinction introduced by Habermas in his Theory of Communicative Action[13] provides a framework for explaining this movement. Habermas analyzes markets and administrations as "systems" that coordinate social action objectively. The potentially conflicting intentions of the individuals are harmonized not by explicit agreements, but by an institutional framework and simple procedural rules. Buyers and sellers, for example, act together on the market for their own mutual benefit without the need for subjective agreement beyond recognition of the forms of exchange such as price, purchase, and sale. In contrast, the lifeworld consists of communicating subjects whose action is coordinated by mutual understanding of a wide variety of elaborate social codes and meanings. Production is organized primarily through the system, and social reproduction through the lifeworld. The dystopian critique of modernity can be reformulated on these terms as the growing predominance of system over lifeworld, with potentially disastrous consequences for social cohesion and the survival of individuality.

Habermas's schema has some limitations for our purposes. He leaves out technology, although it too coordinates action objectively. He treats his concept of system as a pure analytic category, without recognizing its functional role in actual social life. "Systems thinking" is not the exclusive prerogative of the social critic, but rather grows out of the actual experience of managing modern social organizations. Finally, he tends to hypostatize system and lifeworld as separate institutional spheres, which obscures their complex intertwining in actual social life. The lifeworld perspective is brought to bear on systems by those enrolled within them. This is not an analytic error but a reflection of the way in which alienation is lived and to some extent masked, reduced, and resisted by subordinate technical actors.

Let me offer an example of this from the realm of technology. The telephone network is a system in Habermas's sense, managed in accordance with administrative rationality and distributed on a market. Yet the activities the telephone network supports are essentially communicative and the telephone takes on, accordingly, a meaning and a series of connotations in the lifeworld having to do with intimacy, human contact, security, and so on. The telephone is not merely instrumental to these lifeworldly ends, it belongs to the lifeworld itself as a richly signified artifact. This is more than a matter of subjective· associations since it affects the evolution and design of the network, which cannot be understood on pure system terms. The intertwining of power and meaning exemplified by the telephone is general in modern societies.

Michel de Certeau offers an interesting account of the tensions between systems and their subjects which is helpful in articulating this entanglement of apparent opposites.[14] He contrasts the "strategies" of the managers of modern institutions with the "tactics" of their subordinates. The managers act out of a stable power base with a long time horizon, while ordinary people improvise micropolitical resistances within that framework. These two standpoints correspond roughly with system and lifeworld intertwined as I have suggested here. The strategic standpoint privileges control and efficiency while the tactical standpoint gives meaning to the flow of experience shaped by strategies. In the everyday lifeworld masses of individuals improvise and resist as they come up against the limitations of the technical systems in which they are enrolled. These resistances influence the future design of the systems and their products.

This approach to the relation of system and lifeworld, strategies and tactics goes beyond both dystopian condemnation and the posthumanist acceptance of technology. Dystopianism adopts the strategic standpoint on technology while condemning it. Technology is conceived exclusively as a system of control and its role in the lifeworld is overlooked. The introduction of a lifeworld perspective into the study of technological society completes the picture sketched by posthumanist analysis of technical networks. The contradiction between technology as system and lifeworld that is part and parcel of a technologically advanced society explains the rise of struggles in the emerging technical public sphere.

The utopian and dystopian visions of the late nineteenth and early twentieth centuries were attempts to understand the fate of humanity in a radically new kind of society in which most social relations are technically mediated. The hope that such mediation would enrich humanity while sparing human beings themselves was disappointed. The extension of technical control overtakes the controllers beyond a point we have long since reached. But the dystopians did not anticipate that once inside the machine, human beings would gain new powers they would use to change the system that dominates them. We can observe the faint beginnings of such a politics of technology today. How far it will be able to develop is less a matter for prediction than for practice.

Notes

This chapter is a revised version of a contribution to the "International Symposium on The Twentieth Century: Dreams and Realities," Hitotsubashi University, December 2, 2000.

1. Marshall McLuhan, *Understanding Media: The Extensions of Man* (New York: McGraw-Hill, 1964), 46.

2. This projection implies the application of a primitive notion of marginal utility under conditions of income equality, which is not, be it noted, a Marxian desideratum. The varying preference for leisure remains as a basis for the rational allocation of labor. Unfortunately, this appears to create a vicious circle: the least popular jobs would have the shortest hours, requiring the recruitment of a large number of workers who would have to be offered still shorter hours at the margin, and so on ad infinitum. Still, it is a nice try for 1888!

3. Quoted in Tom Rockmore, *On Heidegger's Nazism and Philosophy* (Berkeley: University of California Press, 1992), 241.

4. Herbert Marcuse, *An Essay on Liberation* (Boston: Beacon, 1969), 19.

5. Andrew Feenberg and Jim Freedman, *When Poetry Ruled the Streets: The French May Events of 1968* (Albany: State University of New York Press, 2001), 84.

6. Bruno Latour, *Nous n'avons jamais été modernes* (Paris: La Découverte, 1991).

7. Donna Haraway, "The Cyborg Manifesto," in *Simians, Cyborgs, and Women: The Reinvention of Nature* (New York: Routledge, 1991).

8. For a detailed case study of this transformation in the first successful domestic computer network, the French Minitel system, see Andrew Feenberg. *Alternative Modernity* (Los Angeles and Berkeley: University of California Press), chap. 7.

9. Andrew Feenberg, "The Written World," in *Mindweave: Communication, Computers, and Distance Education*, ed. A. Kaye and R. Mason (Oxford: Pergamon Press, 1989).

10. This is what is wrong with the many polemics against information age hype. When philosophers believe they need not discuss the reality of the technologies they study, but merely respond to the silliest prophecies of enthusiasts, they fail us. As the straw men hit the ground bleeding, we are left wondering what, after all, *is* actually happening. For a more measured approach, see Maria Bakardjieva and Andrew Feenberg, "Community Technology and Democratic Rationalization," in *The Information Society* (forthcoming 2002).

11. John Dewey, *The Public and Its Problems* (Athens, Ohio: Swallow Press, 1980), 126.

12. Andrew Feenberg, *Transforming Technology: A Critical Theory Revisited.* (New York: Oxford University Press, 2002), 74–80.

13. Jürgen Habermas, *The Theory of Communicative Action: Lifeworld and System: A Critique of Functionalist Reason*, trans. T. McCarthy (Boston: Beacon, 1984, 1987).

14. Michel de Certeau, *L'Invention du Quotidien* (Paris: UGE, 1980).

Part Two

❖ ❖ ❖

Humanity

7

The Problem with "The Problem of Technology"

Arthur M. Melzer

> It is not an exaggeration to say that the question of
> technique has now become that of the destiny
> of man and of his culture in general.
>
> —Nicholas Berdyaev

T hinking is called into existence by problems. The thought not only of individuals but of whole eras thus tends to organize around some prevailing conception of the deepest question. In our own age, this catalytic, crystallizing question seems, more and more, to be that of technology. All the great issues of morality and politics, formerly addressed on their own or in a religious context, are now placed in the larger context of technology, which suddenly seems the most powerful and pervasive force, the most ultimate phenomenon.

The clearest evidence for this change is the remarkable convergence that has occurred in the last generation between theorists of the radical Right and Left: both now regard technology not merely as problematic but as our deepest problem. On what may be called the romantic and existentialist Right, there has of course been a long tradition of hostility to machine technology as destructive of man's higher, aesthetic and spiritual nature. But in the thought of Heidegger this tendency is deepened and placed at the center: technology, properly understood, is the most fundamental as well as most problematic characteristic of human existence in our age and, at least incipiently, of Western thought and Being since its Platonic origins.[1]

On the Left, the dominant, Marxist view—which arose by repudiating an earlier tradition opposed to industrialization as inherently exploitative—attributed all exploitation to the mode rather than the means of production, to the bourgeois ownership of technology rather than to technology itself. By thus shifting all blame for injustice onto capitalism, Marx not only exculpated technology but was able to embrace it as the ultimate savior and liberator of mankind—a view that triumphed because it was more consistent with left-wing progressivism.[2] But in the last sixty years the thought of the Frankfurt School and its spinoffs in the "New Left" have produced a veritable revolution on the Left, whereby what formerly was embraced as a savior has now been revealed as the devil itself. Today on the Left, no less than on the Right, technology or "instrumental rationality" is seen as the root phenomenon of modernity and as the true source of all our evils.[3]

The emergence of technology not only as a problem but as the organizing problem of our times can also be seen in other, less theoretical phenomena. In contemporary popular culture, for example, what is the largest, weightiest, most "philosophical" question that it is possible to ask? Not if we will be saved; not if there will be a transfiguring, political revolution; but whether we will "destroy the planet." Especially for the college-aged, that is the truly serious question, which quiets the unruly crowd and seems to place everything in its proper perspective.

To exaggerate only slightly, one could elaborate the apocalyptic claims implicit within this conception by speaking of "three ages of man." In the first, metaphysical age, the question that organized human thinking was that of man's relation to the divine or transcendent order. In the second, humanistic period, the issues of humanity's this-worldly security and comfort, and the problem of man's exploitation of man (the "social question") moved to the center. In our own age, the third, while these other problems have not disappeared, the crucial issue has become humankind's relation to the rest of nature, understood now not as transcendent but as something vulnerable and subject to our control, for which we must take responsibility. This is the age of the problem of technology.

Certain as it is, however, that the focus of human thinking has changed radically over time, that fact alone is obviously no proof that it ought so to change. The human diet has also varied over time, which does not prove that some foods do not nourish better than others. It is at least possible that the deepest questions of human existence are fixed—rooted in an unchanging "human condition"—however neglected or variously distorted they may be in different ages. Thus, in knowing ourselves to view the world from a new and altered perspective—that of the question of technology—it is still incumbent on us to wonder how this change came about and to raise the question of its value or legitimacy. This is of course a monumental undertaking. What follows is merely an attempt to survey the terrain, some familiar and some new.

Our Uniquely Technological Age

It is an obvious fact that neither technology nor the question concerning its value is new to the modern period. Indeed, human life as such is so inseparable from technology that *homo sapiens* is commonly distinguished as "the tool-making animal," just as the stages of human civilization are differentiated in terms of the tools men have actually made. The pervasive effect of technology is visible even in the most primitive societies, especially if one considers not only simple tools and weapons, but also ritual art, body painting and surgery, fermented drinks, and so forth. And no recent technological innovation has been so momentous in its consequences as the agricultural revolution.

As for the awareness of technological change as a problem, one might point to the fierce resistance of hunters and nomads to agriculture, which easily rivals the more recent hatred of peasants, artisans, and intellectuals for the machine. On a more reflective level, the stories of Babel and Prometheus suffice to show that some kind of "problem of technology" has been on human minds for a very long time.

And yet clearly there is something distinctive about the modern period. We not only have more technologies, but somehow we are more technological. One senses that the phenomenon of human artifice, which is in some sense universal, has come into its own so radically and pervades human existence so thoroughly, that it now seems to form the horizon within which everything else finds its place. If one could clearly spell out the ways in which this is true, one might see why the problem of technology has taken center stage today, as well as begin to evaluate whether it has a right to that place. Thus, we need to understand two interrelated things: how is modern technology unique and how is modernity uniquely technological?

I will begin with a brief survey of more or less familiar characterizations of technology and of its role in the modern age before turning to the two factors that seem to me at once most important and least adequately understood—the modern belief in progress and disbelief in happiness.[4]

THE GREATER EXTENT OF MODERN TECHNOLOGY

The most obvious characteristic distinguishing modern technology from its earlier forms is simply that there is such a great deal more of it. Of all the objects with which one routinely comes into contact today, a remarkably high and growing percentage are artificial. These technological objects, of which there are more, are also more technological—being artifacts not merely in form, such as an earthenware pot, but also in their very matter. "Plastic" denotes a new level of artificiality. Furthermore, they are machines and not merely tools, for even their motion or power is artificial.[5] And increasingly they are "smart" machines, which replace the power of the human mind as well as of the body.

They are further distinguished by the involvement of new, unanticipated physical forces or effects, such as nuclear energy or laser optics, which are wholly alien to our natural experience of the world. They stand at a greater remove also from our natural productive faculties, being the product not of human hands directly but of machines that are the product of still other machines, and so forth. Finally, they are technological not merely as means for the provision of natural or at least accustomed goods, but exist for the service of ever more artificial ends. The purpose of a television set, for example, cannot be fully understood except in terms of a world of needs as artificial as the television itself.

It is not mere artificiality, however, that distinguishes the products of modern technology, but also ever increasing automaticity, uniformity, and disposability. These are the qualities that give rise to the unique sense of sterility, the existential disengagement and alienation from the world of objects, that seems to be a hallmark of modern experience.[6]

NEW RELATIONS OF DEPENDENCE

Two further characteristics of the modern world follow directly from the great extent of technological development. Our age lives in a condition of utter and unprecedented dependence upon its technology. In the absence of the artificial world we have created for ourselves, very little of modern life—of what we have and do and are—would remain. For example, the prosperity that has made possible mass education and thus mass democracy would disappear. Indeed, perhaps 85 percent of the world's present population, which has increased almost eightfold since the start of the Industrial Revolution, would be insupportable.[7]

Just as striking, however, is the degree to which the rest of nature has become dependent on us. Due to the vast scale of technological development, it is now up to us whether other species will survive or disappear, whether the atmosphere and climate will continue unchanged, whether the planet will remain supportive of human and other forms of life. In all past ages, it was believed that, however great man's evil and foolishness, his sins only fall back on himself, leaving the larger order of nature to go its way serenely unchanged, a palpable symbol of man's pettiness and his need to conform to the order that transcends him. In fact, it still remains true, if no longer so palpable, that man has no power over the basic laws of nature or over the ultimate finitude of human life and the cosmic conditions that support it. Nevertheless, it is difficult to escape the conclusion that something of a revolution has occurred in the structure of the human condition as a result of modern technology. All future thinking about morality and the human good must include the new fact that men have largely succeeded in becoming the "masters and owners of nature," as Descartes had urged, and that this ownership brings with it new responsibilities. The classical virtue of moderation is as crucial today as it was in the past, but henceforth it will be grounded less on nature's overawing permanence than on its vulnerability, less on man's smallness than on his excessive and dangerous technological power.[8]

THE GREATER SCOPE OF MODERN TECHNOLOGY

It is not only the extensive material development of the modern world that makes it uniquely "technological," but also the spread of this phenomenon to other areas of life formerly untouched by it. Beyond physical instruments and machines, there is something that may be called the technological "attitude," or "way of thinking," or even "posture toward Being": a nonspecific, but generally utilitarian understanding of ends, a primary focus on means and power, the restriction of reason to instrumental rationality—the methodical pursuit of the one, maximally efficient way of doing each thing—the faith in human self-reliance and control, the belief in the superiority of the artificial to the natural and of the mechanical to the human, and the view that everything man encounters in nature or history is only raw material which he is free to transform for his own purposes. As has been shown with particular force in the work of Jacques Ellul, this attitude, which used to be found only in a few, narrowly circumscribed areas of life, now seems to dominate virtually all.[9]

The whole idea of "revolution," for example, which stands behind most modern political thinking, imitates man's conquest of physical nature. As Robespierre urges, "Everything has changed in the physical order; everything must change in the moral and political order. Half the revolution of the world is already accomplished; the other half must be achieved."[10] All liberal political systems in particular are decisively technological in their theory and practice. In such thinkers as Locke, Montesquieu, and the American Founders, the state is seen as an artificial human invention, needed to correct the gross defects of nature, and as something fundamentally instrumental, a mere means for the provision of private ends. The citizens, as well, are conceived as "individuals," that is, independent human atoms subject to internal drives and external manipulation. The machine of state that organizes them is to be constructed, not in imitation of some natural or divine pattern, but in the most effective and efficient way. Thus, great reliance is placed on constitutional mechanisms and techniques—separation of powers, checks and balances, representation, party government, federalism, extending the sphere of the free market—rather than on human qualities such as statesmanship or moderation or justice. As stated (perhaps overstated) in the title of a recent book on the American Constitution, the goal was to build "A Machine That Would Go of Itself."[11]

Similarly, in the sub-political realm of work and economics, we have created an instrumentally rationalized world of bureaucracies and assembly lines, of specialized men and standardized products. Even on the level of leisure and amusement, one sees a similar tendency, perhaps less planned but no less obvious, toward the triumph of technique. Indeed, it is difficult to think of any human activity, grand or trivial, for which there does not now exist a specialized method, therapy, or system, whether for eating or fighting, for dating or proselytizing, for relaxing or motivating, for making cars or making love. Time itself has been captured by clocks and schedules, space by urban grids and highways, value by dollars and cents. Gradually, all alternative principles of experience and choice—piety, morality, aesthetics, custom, instinct—have been, to paraphrase Marx, drowned in the icy waters of technical calculation.

THE NEW SCIENTIFIC BASIS

A still more distinctive and profound characteristic of contemporary technology is its intimate connection with modern, theoretical science. In prior ages, each of the different arts was based on a relatively separate and self-contained body of practical "know-how" or crafts-knowledge, acquired through trial and error and transmitted through tradition. Indeed, the practical arts were necessarily separate from "science" or "theory" owing both to the latter's motive—the idle, disinterested contemplation of what is—and to its object—the realm of changeless, eternal beings that cannot be other than they are. A revolution in human thinking was required to produce the situation that today we take for granted: the derivation of the crafts from natural philosophy.

This revolution has had, as its most obvious consequence, the production of remarkable new craft-objects, which are able to harness forces of nature unsuspected by the prescientific, artisanal world. A deeper if less visible consequence, however, is the fundamental transformation of our attitude toward these practical objects, which also now carry with them all the splendor and dignity of "theory." Modern science constitutes the dominant theory of reality, the highest intellectual authority of our age, much as the Bible did in former times. Today, technological objects have therefore taken on something of the character of sacred objects or divine gifts. They are seen no longer as merely the fruit of some particular craft or artisan but as a remarkable demonstration and consequence of our successful contact with the deepest ground of reality, a sort of message or gift from on high. In other words, prior to our own, there had never been a world in which the most ordinary, mundane objects were so directly and convincingly linked to the most ultimate principles. The great veneration we pay to high technology, and its peculiar prestige in the world—its use, for instance, as the dominant measure of a society's degree of civilization—would seem to derive less from the material comforts it provides than from this: that it constitutes an indubitable revelation and local incarnation of the deepest forces of the universe.

The modern convergence of the arts and natural philosophy, however, involves the transformation of the latter no less than the former; and this change is responsible, on a still deeper level, for the uniquely technological character of our age. At the dawn of the modern era, natural philosophy was transformed into "modern science," which is, in itself, even prior to all practical applications, essentially technological. The whole possibility of "applied science" is implicit from the very start in the structure of "theoretical" modern science. The latter is a fundamentally technological way of knowing.

One sees this, in the first place, in the transformation of the motive or end of philosophy, both natural and moral, enunciated by such thinkers as Bacon, Hobbes, and Descartes. The explicit purpose of "theory" is no longer disinterested contemplation but power. Speculative metaphysics falls into disrepute as the goal shifts from understanding and admiring the world to controlling it.[12] The fact that it was really not until the nineteenth century that theoretical science actually began to yield practical fruits only makes more impressive these early declarations of technological intent.

In its method, as well as its motive, the new science was inherently technological. This is visible, to begin with, in the very idea of "method," which now comes to the fore—the idea that the natural workings of the human mind are not to be trusted and must be replaced with a specialized, artificial technique for reasoning. Modern science does not, like earlier philosophy, grow out of and perfect common sense, raising it to "wisdom," but rather starts afresh, building a separate edifice of technical, "scientific" knowledge, which sits alongside and in permanent tension with our natural way of knowing. Methodological science is a new thought technology.

The content of modern method, no less than its form, is also technological. Whether because of their new, practical motive, or a change in metaphysics, or their new epistemology (this controversy need not be settled here), the very questions that such thinkers as Galileo, Bacon, and Descartes were asking, their whole conception of what would count as "understanding" nature, predisposed their thinking in a technological direction. Prior to all investigation, they tended to dismiss the older questions of "Why" and "What" for the problem of "How." Henceforth, to understand nature would mean to know how it worked.

The precise method to be used in such research—which emerged not so much in any one of these thinkers as from their combined effect— had three central characteristics, each in its own way technological: it was analytical, mathematical, and experimental. By proceeding through "analysis," by resolving every problem or phenomenon into (purportedly) more fundamental independent parts, scientific investigation rejected from the start—in its very method—the possibility that things must be understood as integral wholes (essences, formal causes, organisms) or understood in terms of some larger, overarching order or purpose (final causes). These excluded possibilities—that the world naturally tends toward the realization of certain structures or certain ends—would have placed a limit on both the legitimacy and possibility of the human manipulation of nature. Thus, modern science's analytical or reductionist method, by denying the integrity of natural wholes, by requiring us to break nature down (at least in thought) in order to understand it, is inherently technological.

This point can be strengthened by considering an obvious objection. The ancient Epicureans, with their metaphysical materialism, also rejected formal and final causes, and yet did not favor conquering nature but rather living in accordance with it. The difference derives not only from Epicurean moral doctrine but also from their belief in the irreducible heterogeneity of qualitatively distinct atoms, which puts strict limits on the malleability of nature. By contrast, the modern discovery of consciousness, or of "subjectivity," or of the primary/secondary distinction placed man outside of nature, at least to the extent of being the repository of all secondary qualities; it thus drained external reality not only of noetic but of all qualitative or sensuous heterogeneity, leaving only an abstract and purely quantitative world of homogeneous matter. Consequently, whereas Lucretius explained the different qualities of things in terms of the inherent properties of their atoms, modern science sought to explain everything in terms of the different configurations or combinations of uniform matter. And since these configurations, once understood, might be recreated or transformed by man, this more radical reductionism or "analysis" opened a whole new vista on the manipulation of nature.

In breaking completely with the "qualitative" world and reducing everything to quantity and relation, modern science in fact went beyond mere analysis to a unique mathematicization of physics. Again, it is true that Pythagoras and, on some interpretations, Plato also saw mathematics as the key to the universe, but by this they meant that the cosmic whole had a permanent, static structure or form based on certain mathematical relationships. Modern science, by contrast, holds that the universe neither possesses nor tends toward fixed structures of any kind, but is in constant, aimless flux. It is to the laws of this flux, and not to any static order, that modern science applies its mathematics. And whereas "mathematical physics" in the Pythagorean sense is clearly antitechnological and harmonistic, in the modern sense it is the very opposite. It means that, on the one hand, the world is vastly more changeable or malleable than at first it appears to be, and yet, on the other, that its constant change is not some incomprehensible, Heraclitean flux but rather a precise, determinate, lawful sequence of cause and effect, susceptible of mathematical prediction and thus human control.

As analytical and mathematical, scientific method grounds the possibility of controlling nature; as "experimental" it proclaims the necessity for it. Modern science is built not on idle, speculative reason, nor even on empirical observation as such, but on the active manipulation of nature. The assumption is that nature, which is not ordered so as to serve man's basic needs, is also not organized to serve his need to know. There is nothing privileged about nature's spontaneous self-presentation, nothing "more natural" about the forms it happens to assume in our particular environment. To extract its secrets, therefore, it is necessary to abandon contemplative detachment and passivity, to dissect, probe, and transform nature, breaking it down through a kind of "analysis in action," in order to isolate its component parts for observation and measurement. In modern method, knowing is strictly inseparable from acting and manipulating.

As our knowledge has progressed, the reliance of theory on action has grown ever greater. It was not very obtrusive when all that science required was an inclined plane or simple telescope, but today when vast feats of invention and engineering are needed—such as huge accelerators and interplanetary probes—one is struck by the utter dependence of theoretical knowledge on technological mastery. Modern science lives by the principle: one cannot understand the world without changing it.

In sum, the whole possibility of "applied science" and of the technological mastery of nature did not just happen to turn up as an innocent finding of theoretical science, an unanticipated discovery. It is not at all posterior to or separate from theoretical science, but rather implicit from the very start in the latter's conception of "understanding" and of the proper method—analytical, mathematical, and experimental—needed to pursue it. Modern science is intrinsically—intentionally and methodologically—technological.[13]

The modern convergence of the crafts and natural philosophy, then, does indeed involve the transformation of the latter no less than the former. And to the extent that modern science has become the reigning theoretical authority and the model for all genuine knowledge, it is possible to say that we are all prisoners of a technological consciousness." In its every dealing with the world, the modern mind tends to a kind of thinking that is technological in form, content, and motive.

TECHNOLOGY BOUND AND UNBOUND: THE IDEA OF TRADITION AND THE IDEA OF PROGRESS

Modern technology is unique, we have seen, by virtue of its more intensive and extensive development, its spread to every realm and aspect of life, its creation of a new dependence of man on technology and of nature on man, and its intimate, reciprocal relation to natural philosophy. Yet, to understand the full meaning of technology and especially its constitutive role in the moral-psychology of modern life, it is necessary to go beyond the more or less familiar themes just discussed to two further points, perhaps no less familiar, but less adequately understood. These are the modern belief in progress and disbelief in happiness; or, in terms less misleading in their simplicity, the "futurity" and "negative orientation" of modern life.

In the first place, what we mean by "technology" is ultimately unintelligible without a proper conception of its link to that larger movement of which modern science is still only a part: the faith in progress, the commitment to humanity's gradual self-liberation through its progressive appropriation of the natural and historical worlds. In the premodern era, after all, the various arts were in no sense conceived as a single historical force engaged in a long-term battle for the human conquest of nature. But this thought stands at the very core of the modern concept of "technology."

Prior to the "age of progress," the arts were not seen as part of some great journey of mankind—a historical "project" or "movement"—but simply as individual crafts addressed to specific and limited needs, and as essentially set in their ways, even if susceptible of gradual improvement. Moreover, far from glorifying man's freedom and power, the arts anxiously endeavored—for reasons to be considered momentarily—to obscure this as much as possible, attributing their origins to the gods, taking their models from nature, hedging themselves in with a hundred arbitrary rules and traditions, and embedding their practices in religious ritual and elaborate social structures (trade secrets, apprenticeships, guilds, and so forth). In short, in premodern or traditional society, the arts tread quietly, outside the circle of men's deepest cares and loyalties, and indeed in bondage to them. This was the age of technology bound.

All of this changed with the rise of humanism and the idea of progress. All external and limiting conditions on the arts were gradually cast off, man became a conscious creator, "method" came to the fore as an art of discovery, "inventor" became a job, research and development were institutionalized, and the advance of human control became a deliberate, universal, open-ended project. It is only this revolutionary change that gave birth to "technology" properly so-called, in which every particular art is experienced as part of the larger movement of all humanity toward freedom and mastery.

This great historical transformation is of course familiar enough, but its full significance for the uniqueness of modern technology cannot be seen if it is understood in the conventional way, as a triumph over the simple cowardice and superstition of the past by the *sapere aude* of the Enlightenment. The fact is that all premodern societies tend to be deeply suspicious of technological innovation and of science; and while this tendency may stem in part from mere fear, prejudice, and ignorance, it is also a necessary concomitant of the whole structure and life orientation of such societies—of their rootedness in tradition. Only through the wholesale reordering of this comprehensive moral-political foundation could the ideal of progress emerge and, with it, technology properly so-called. Thus, this is the change we need to understand.

In all societies untouched by "modernization" or "Westernization," the foundation of the social world—the legitimacy of the government, the respectability of the social hierarchy, the authority of prevailing mores—is rooted in ancestral custom or tradition. They are "traditional societies." But how does such a society work? What is the source of the power of tradition over human minds and hearts? This power may be said to rest, I believe, on three distinct forces—each of which will turn out to be in necessary tension with technological innovation.

The force of tradition derives, in the first place, from that of habit: we naturally develop a loyalty, fondness, and inclination for that to which, over a long period of time, we have become "accustomed." But the importance of habituation leads to an essential tension between custom or law and the arts, as Aristotle states in his classic discussion of whether laws should be as changeable as the arts: "The argument from the example of the arts is false. Change in an art is not like change in law; for law has no strength with respect to obedience apart from habit, and this is not created except over a period of time. Hence the easy alteration of existing laws in favor of new and different ones weakens the power of law itself" (*Politics,* 1269a19–24). Since habituation requires changelessness, a healthy traditional society must be relatively static; and therefore, it must necessarily be suspicious of the example and influence of the arts, with their reformist and innovative tendencies.

The second source of the power of custom is the communal consensus or public opinion in which it is embedded. The widespread judgment of society, and especially of its oldest members, carries—not unreasonably—a great authority, which the individual naturally tends to trust and respect in preference to his own limited powers of discernment. This rational deference to consensus is further supported and strengthened by the censorious pressure of public opinion, which rewards agreement and harshly punishes dissent. Custom, in other words, draws upon a certain natural inclination within groups toward intolerance, closed-mindedness, and conformity—an inclination that modern culture deplores as senseless, but which would seem to be essential (within measure) to the self-preservation of a traditional society, since its authority and world-view are supported, not by rational demonstration, but only by the brute fact of actual agreement. Outside the communal consensus lies an abyss; and sensing this, people naturally hate what is strange. It is perhaps not impossible for such a society to maintain its self-certainty in the face of fundamental dissent or diversity, but it is always an ordeal. Consequently, reason and the arts, while certainly necessary to such societies, must always remain essentially suspect owing to their tendencies to skepticism and innovation, and their epistemological stance outside of consensual authority.

Finally, tradition exercises its power over the minds of men because it is old or ancestral. Things very old, as those very large, naturally inspire awe, since they dwarf our own limited extension in time and space. We also revere ancient customs because they represent the consensus of long generations of men, as well as the product of a searching test of time.

Without further support, however, these considerations could not hold out against the debunking of tradition contained in the following reflection: "The first [human beings], whether they were earthborn or preserved from a cataclysm, are likely to have been similar to average or even simpleminded persons [today] . . . so it would be odd to abide by the opinions they hold" (*Politics*, 1269a3–8).

What truly vindicates the reverence for tradition, then, and constitutes its deepest ground, is the belief—essentially religious—that our ancestors were vastly superior to ourselves. There is a logic to such a view. Just as parents are stronger and wiser than children (who also owe an infinite debt to their progenitors), so our ancestors were greater than we. What comes earlier must be superior to what comes later, since the former has created or begotten the latter. Indeed, thinking one's way back to the very beginning, one realizes that the divine beings of the first age, who, without being the children of any earlier generation, generated the world itself, must have been absolutely the most powerful. In our own, late, and feeble age, this power has either diminished or withdrawn from the world, since changes of such magnitude no longer occur. But our earliest ancestors were close to this power, at the high-water mark of the world, a golden age when heroes strode the earth and men conversed with gods. The ancient traditions, then, handed down to us from our ancestors are to be cherished and revered. They come from a god-like source far greater and wiser than ourselves.

This foundational belief in the inherent superiority of the past to the present represents the third and greatest source of inescapable tension between the arts and traditional society. Technological innovation threatens not only the fixity needed both for habituation and for communal consensus, but also the indispensable attitude of reverence for the ancestral. In other words, what it means, on the deepest level, for a society to be "traditional" is not merely that its subjects happen to follow old customs, but that, on some level, they understand themselves to be fallen from the originary fullness of the world; that they live in remembrance, repentance, and hunger for the past; and that they cleave to tradition as their sole, tenuous link to the ever-receding, life-giving, ancestral source. In this light, the arts must appear deeply suspect, for novelty is essentially betrayal. And our familiar idea of progress, could it arise, would seem altogether reckless, unholy, and insane.

In sum, in a society constituted on the basis of tradition, the opposition to reason and the arts does not derive from mere timidity or ignorance but reflects a permanent structural conflict between this open, dynamic, progressive aspect of human nature and the, as it were, ontological rootedness of communal life in the ancestral past. To employ a Nietzschean idiom, there exists here a fundamental tension between thought and life. This is not to say, obviously, that the arts cannot exist in a traditional society, but rather that they do so precariously and only by virtue of the elaborate accommodations described above.[14] They are eternally gnawed upon by anxiety and guilt, as Prometheus would discover.

It follows, conversely, that in order for technology finally to be unbound, as in the modern era—in order for life actually to embrace reason, and humanity the ideal of progress—it is first necessary to supply an entirely new foundation for morality and politics. That foundation therefore constitutes the indispensable—but long-forgotten—ground of the modern turn to technology.

It was with this thought in mind that the Enlightenment consciously endeavored to subvert traditional society, to liberate men from all customary and religious authority, and to reconstitute the state on the basis of the free consent and enlightened self-interest of the resulting autonomous, rationalistic, and future-directed individual. In this new world, obedience would result not from the static structure of habit, consensus, and reverence for the past, but from the dynamics of the desire to get ahead—not from changelessness but precisely from the collective pursuit of change. Similarly, the legitimacy of the political hierarchy would no longer be based on some delicate communal consensus regarding ethics and religion but on the morally neutral principle of contract or consent. In such a society, which thrives on innovation, diversity, skepticism, and materialism—on everything most lethal to a traditional community—the technological impulse could at last be fully unbound. For the first time in history, it would no longer exist in fundamental tension with man's primary loyalties.

Indeed, technology now becomes not only permissible but, in the same moment, necessary. For, the social bond of enlightened self-interest—which, replacing that of tradition, permits the unleashing of technology—is not in fact a bond at all, rather than a source of conflict only insofar as the economy expands over the long run. And, in the absence of imperialism, the economy can expand steadily only if there are continual gains in productivity stemming from the advance of technology. In other words, in the modern, liberal-democratic, mass nation-state—the ultimate result of the Enlightenment's assault on traditional society—the diverse, deracinated, self-interested individuals, sharing no bond of "race, creed, or national origin," can live in harmony only through a certain hostility toward nature. We liberated men and women of the free society are united only by sharing the spoils of our common technological imperialism.

But there is still more. In the reconstitution of life engineered by the Enlightenment, technology becomes not merely released from its age-old bondage to tradition, and not merely an indispensable economic condition of the system that releases it: it is moved from the periphery to the very heart of what people live for. As the primary vehicle or expression of progress, it replaces tradition as the moral center of men's lives.

In this new society based on self-interest, after all, there is still idealism and morality. The idealistic citizen lives for the "getting ahead" not only of himself but also of larger communities. In the first place, this means his family—that is, the new, future-oriented, child-centered family. He lives, not for his ancestors, but his children, whom he considers in some sense his superiors. Beyond that, he seeks to "be useful" and to "make a contribution"—but to what? Again, not to the ancestors and not even primarily to the present community, but to the future. He feels the need to have "a cause"; he wants to "help build a better world." This is now what it means to be "idealistic." As always, it means devotion to a larger whole, but the whole is now not something extending back in time but forward—no longer a fatherland but a "movement."

The moral life necessarily takes the form of membership in a movement, and that means it is rooted in the future and in the idea of progress. For the idealistic modern citizen, who he is is, where he is going—and not where he has come from. He lives, not in remembrance, but in activist expectation, and understands himself, not as fallen from some golden age of primordial fullness, but as part of the progressive march of civilization beyond the barbaric past (the "state of nature") and toward a better future for all. To be sure, powerful vestiges of tradition remain, as of course do obligations to the present community. But these ties must always justify themselves as forces for progress. Modern patriotism loses legitimacy if one's nation is shown to be on the wrong side of history. In a manner unintelligible to the traditional citizen, the future of humanity now constitutes the highest good and the ultimate homeland of men's hopes and loyalties.

In other words, the replacement of the idea of tradition with that of progress means far more than that people now tend to the belief that things will improve. It is a fundamental reversal of life orientation, in which the source or fullness of being is relocated from past to future. Human lives now derive their meaning and weight not by rootedness in the ancestral but by being the wave of the future. Modern man is a forward-looking, upwardly mobile, futural being whose profoundest suffering is no longer the guilty sense of fallenness or estrangement from the ancestral source, but rather the fear of failure and the terror of being left out or behind.

In this transformed world, technology inevitably becomes the guiding and inspiriting force. We derive our sense of right from feeling that we are in touch with the future; and the most palpable sign of that contact is technology, which is, as it were, the most futural thing. It offers the surest proof that there has been progress and that there will be more; it is the clearest candidate for a world-historical force. Indeed, today, when one tries to form a picture of the distant future, one feels unsure about political and economic arrangements, but one always imagines it more technologically advanced. In our dynamic world, that is our one certainty: the future belongs to technology. That is why nothing is more deeply demoralizing and subversive in the present age than shoddy and outdated technology, as the experiences of Eastern Europe would seem to show. The good is the "advanced." In sum, in this new, future-directed world, it is now tradition that is fundamentally suspect and in tension with "life"; and conversely, it is human art, reconceived as "technology"—a unitary, ever-advancing movement for human mastery—that stands at the center of life, the true spring of human vitality.

There is a common tendency to think of technology as a morally neutral and universal phenomenon whose modern development simply resulted from the "advance of civilization"—the gradual accumulation of discoveries, spurred on by curiosity and material need, while resisted by human timidity and superstition. But the true character of technology, as indeed of modernity itself, cannot be grasped unless one sees that the technological turn was rather a culturally singular event made possible only by a unique moral, political, and, as it were, existential revolution that, overturning the entire traditional order, released the arts from their heretofore necessary bondage and, what is more, placed them at the heart, not only of the political-economic system, but of the whole purposive structure of modern life.

THE REJECTION OF HAPPINESS AND THE "NEGATIVE ORIENTATION" IN LIFE

Closely connected to the progressive, futural, movement-oriented character of contemporary existence is our typically modern skepticism concerning the possibility of rest, happiness, or a *summum bonum*. This skepticism together with its result—the turn to a "negative orientation" in life, which seeks the avoidance of evil rather than the attainment of good—constitutes a second crucial, but neglected element in the rise of modern technology.

One can see this most clearly as follows. In the foregoing discussion of the transition from traditional to progressive society, a middle term could be said to have been left out. Human life may be anchored in the ancestral past or eschatological future, but also in that which is timelessly present: a rational, natural standard of the human good—of happiness or the *summum bonum*—as elaborated, for example, by Aristotle in the *Ethics*, and indeed by virtually all the competing sects of classical philosophy.

If one evaluates technology from this alternative standpoint—neither tradition nor progress, but the natural human good—one comes across the remarkable fact that virtually all of the ancient schools—Platonists as well as Aristotelians, Stoics no less than Epicureans—agreed in opposing the unregulated development of the arts, the endless pursuit of mastery, and unlimited acquisitiveness. Indeed, two millennia before Rousseau, most of them entertained a "primitivist" skepticism about the real value of civilization and its "progress."[15]

To some extent these attitudes simply reflected an acknowledgment of the requirements of traditional society which we have just considered; for the "Greek enlightenment," in contrast to that of the eighteenth century, never envisaged the wholesale replacement of such a society with one based on reason. But, this aside, their reasoned conception of the good was itself also antitechnological. It was argued that knowledge—the perfection of the mind—must surely have a worthier and more satisfying purpose than finding comforts for the body. Indeed, the intrinsic joys of contemplation soon disgust one with all such vulgar utilitarianism. Even the greatest ancient technologist, Archimedes, held to this view (as reported by Plutarch): "He would not deign to leave behind him any commentary or writing on such subjects; but, repudiating as sordid and ignoble the whole trade of engineering, and every sort of art that lends itself to mere use and profit, he placed his whole affection and ambition in those purer speculations where there can be no reference to the vulgar needs of life."[16]

Even the Epicureans such as Lucretius, who rejected the life of contemplation and identified the good with pleasure, shared this contempt for the arts and acquisition. What, after all, can technology really do for us? The true needs of nature are very few and easily satisfied. Pleasure, like all things, has its natural limits; beyond that, all is vanity and illusion. If one chases after power and possessions in the hope of satisfying all of one's desires, one will only succeed in increasing toil, strife, and the dependence on men and fortune, but not one's contentment; for such desires are infinite and always run ahead of our power to satisfy them. It is better, then, to put desire and power in harmony, not by extending the latter, but by limiting the former—by moderation, the classical virtue.[17]

Of course, the same conclusion is reached if, like the Stoics and others, one identifies the good with moral virtue. In fact, so long as one has any conception of the good or final end, regardless of its content, one will reject as utterly senseless the endless pursuit of means.

It is crucial to note, therefore, that modern thought begins precisely with the rejection of the possibility of happiness or a *summum bonum*. As Hobbes proclaims in a famous passage:

> The felicity of this life, consisteth not in the repose of a mind satisfied. For there is no such finis ultimus, utmost aim, nor summum bonum, greatest good, as is spoken of in the books of the old moral philosophers.... Felicity is a continual progress of the desire, from one object to another; the attaining of the former, being still but the way to the latter.[18]

There is nothing the possession of which brings satisfaction. Hence, there is "felicity," the pleasure attending the ceaseless motion from one temporary goal to another, but no happiness—no contentment, completion, and repose. Life is endless wanting and striving.

But, then, striving for what? If there is nothing that brings satisfaction or happiness, then for what purpose and with what hope do we make something the goal of our striving? According to Hobbes, there can be no positive goal of human striving but only a negative one: the avoidance of evil and especially of the greatest and most comprehensive evil, death. His radical "noneudemonism" entails a fundamental reversal of life orientation: the pursuit of good is replaced with the overcoming of evil, the *summum bonum* of happiness with the *"summum malum"* of death.

But the avoidance of death or the attainment of security is an unfinishable task, and therefore the proper and inevitable character of human desire is this: "a perpetual and restless desire of power after power, that ceaseth only in death."[19] In this way, the pursuit of means and power was liberated from all concrete ends. Through this momentous revolution, technology, already unbound from the fetters of tradition, was cut loose from those of the human good as well—set free to become, for the first time, an independent goal and indeed the central object of philosophy, society, and progress.

What Hobbes wrought has not yet been undone. The negative orientation toward life—founded on the dismissal of happiness—continues to dominate modern culture, which does not hold up for emulation any image of the completed or perfected life. We believe in goods, to be sure, in various pleasures and comforts, but not in a complete good or ultimate purpose that might structure a way of life. In the absence of such a comprehensive positive goal, the lives of individuals and of nations tend to be organized negatively, by the avoidance of death and the derivative evils of war, hunger, poverty, disease, oppression, intolerance, and superstition. These evils, and not some vague talk about happiness, are what we really believe in. We define progress in terms of the distance traveled away from them (or from the "state of nature," "barbarism," the middle ages, the "old country"). Lacking positive goals, knowing what we are against but not what we are for, we tend to cherish opportunity over accomplishment, rights over goods, liberation over consummation, youth over maturity, hope over fulfillment, and, above all, means over ends. Consequently, we embrace technology with an abandon that would not be possible were there clarity regarding the ultimate end. What might be called the "forgetfulness of happiness" thus lies at the core of the technological turn.

Some Problems of Technology

Having examined the character of modern technology and the technological character of modernity—with special emphasis on the futurity and negative orientation of modern life—we are in a position to assess some of the more prominent formulations of technology's "problem." Our foremost concern is to know whether this problem deserves the position it has recently assumed as the central or most fundamental philosophical issue.

THE IRONY OF TECHNOLOGY

All serious conceptions of the problem presuppose some form of the thesis that technology—which seems to be a mere means and thus a derivative phenomenon, subordinate to human choice—has become in fact an independent force that rules its putative master. The seductive irony of this thought has encouraged exaggeration, but it is difficult to deny that there is much truth to it.

First, there is the fact that as soon as one chooses a means, one begins to obey it—to do the myriad things necessary for its acquisition, maintenance, and employment. Think of all the ways we are ruled, for example, by our use of oil as a means. To put it in Marxist terminology, the "means of production" determine the "relations of production"; or, as Rousseau insisted, nothing is more enslaving than the pursuit of power, for, to control one must obey.

Second, every choice of means usually ends up subjecting us to important "unintended consequences" that we did not and would not choose, such as pollution or global warming. One might of course reply to this point, as to the first, that we still remain free in that we can choose or reject the whole package of a given means taken together with all of its requirements and its consequences. But in practice, we seldom learn the full contents of this package for a very long time, and by then the means may be so built in to our world that we hardly remain free to reject it. When the automobile was invented, we had the power but not the knowledge to rule it; later, we acquired the knowledge but lost the power. And that is how it must be with technological progress, which brings a constant stream of new, unknown forces.

Third, we lose our freedom to control a given means not only because it integrates itself into the world but also into ourselves. Objects we first choose as luxuries soon become needs. Similarly, things such as television do not remain outside us, where we are free to judge them, but gradually alter our very tastes and perceptions. In general, as the world becomes more technological, we become more technological—in our sensibilities, our modes of thinking, perhaps in our very posture toward Being. Our judgment is transformed by the very thing we are trying to judge. Thus, toward so pervasive and transforming a force as technology, we are unable to maintain a truly external and free relationship.

Finally, choices regarding technology are not made by single agents in isolation, but in a system with other, competing players. Frequently, we must either develop and adopt new technologies or lose all—militarily or economically—to others who do. It is often the case that no one would choose a given technique for its own sake—dangerous and expensive weaponry or an efficient but environmentally harmful manufacturing process—and yet the system forces all to do so. In a competitive system, inventions rule their makers.

In sum, technology appears to be more than a mere collection of means, subordinate to human choice, but an independent force, with its own logic or destiny to which humanity is compelled to submit. This claim forms the starting point for all conceptions of technology as the most determinative or fundamental phenomenon.

While there is clearly an element of truth to this claim, evident from the sorts of general arguments outlined above, we are perhaps too inclined to think that the issue can be settled on the level of these kinds of arguments. In fact, it would be an enormously complex philosophical-sociological-historical undertaking to try to establish just how independent and controlling a force technology is in practice. Without entering into this question further here, I would simply observe that no one has yet succeeded in answering it convincingly or even in showing how one should go about answering it.[20]

A related question, which remains on our more general level, is what the inner logic or destiny of this more or less independent force might be and how it renders technology the central problem. Returning to the rough typology employed at the beginning, we can distinguish three prevalent answers: a popular or commonsense view, and those of the radical Left and Right.[21]

THE "POPULAR" PROBLEM OF TECHNOLOGY

In the public mind today, technology is seen as a relatively independent force, having as its primary "logic" a simple tendency toward expansion: over time, technology multiplies the kinds and magnitude of human interference with nature. And this is seen to be a problem owing primarily to the corresponding growth of "unintended consequences"—ozone depletion, deforestation, contamination of food, air, and water, and so forth.

But practical dangers of this kind, however urgent they may be, do not lead to the raising of serious theoretical questions about technology as such. There may be some general hand wringing of a philosophical nature, but when people finally set themselves to the task of finding a solution, their primary hopes soon turn back in the direction of new technology (e.g., solar energy). Indeed, for this reason the problem of unintended consequences, so far from undermining the technological project, actually strengthens it, giving people the sense that it is no longer really an option to stop, that now more than ever we need the enlightened technology of tomorrow—in order to save us from the dangers of today's.

Even in their motivation, people who react to problems of this kind remain essentially technological. Now that everyone sees technology as an "independent force" able to do us harm—indeed, a force more alien and harmful than nature itself, which now seems our friend—it inevitably occurs to them to bring it too under human direction and control. No one proposes that we should submit to such historical forces with resignation and acceptance. We must master them. The mainstream movements of environmentalism, antinuclearism, and so forth are simply the modern technological attitude now extended to the force of technology itself.

THE PROBLEM OF TECHNOLOGY ON THE LEFT

On the popular or commonsense level, the problem of technology remains a technological problem. It becomes a political and philosophical

one only on the basis of more determinate views of the destiny of technology, especially as it relates to prevailing political forms. The Left
argues that it is quite impossible to meet the above dangers of unintended consequences within the confines of a capitalist, pluralist, representative, liberal democracy.

First, capitalism is so utterly dependant on technology, the *sine qua
non* of continued growth, and so immersed in its ethos, that it is incapable
of recognizing its dangers, let alone dealing with them effectively. Second,
in a pluralist society, given over to competing material interests, there is no
institution that looks to the good of the whole, nor any interest group or
lobby for the future; in a representative democracy, unborn generations
have no vote.[22] In such a system, where could the political force come
from that would protect the environment or preserve resources for the
future? Thus, only a socialist society—planned, humanistic, and wary of
growth—possesses the ability to tame technology.

But, third, if we do not succeed in thus taming it and freeing our
minds from its mode of thought—instrumental rationality—then we will
be led by its irresistible logic to the ever greater "rationalization" and
organization of society, the transformation of men into tools and cogs,
the loss of all individuality and spontaneity, the rise of technocratic elitism, the strengthening of the state through new technologies of military
force, surveillance, information processing, mass media, and propaganda,
and in sum, the creation of a dehumanized technological despotism.[23]

While for many years it seemed almost axiomatic that capitalism was
in thrall to technology and that the latter was a force hostile to liberty,
recently these views have come to sound strangely old-fashioned and out
of touch. The third argument, in particular, seems difficult to sustain in
light of the undeniable trend in the world toward democracy and, what is
more, the important contributions to this trend made precisely by the new
information and communication technologies. More generally, today the
logic of events seems to be pushing away from social regimentation or
mechanization and toward individual initiative as the source of cooperation
and energy; away from central planning and toward local responsibility and
experimentation; thus also away from socialism—which suddenly seems
the more narrowly technological of the two—and toward free-market
spontaneity. The great temptation today is actually to invert the old left-
wing view and argue that the inner logic of technology leads to democracy
and capitalism. But from this striking shift of perspective a more reasonable
conclusion would be that there is probably no such logic at all, and that,
over the long term, the social and political effect of technology will continue to change in unknowable ways.[24]

The first two arguments also seem to have been refuted by events. It is undeniable that the contemporary environmental movement first arose in the United States and other Western democracies and every year becomes more widespread and powerful. There is a nontrivial sense in which almost everyone today in the United States is an environmentalist, albeit with varying degrees of fervor and influence. So how is this possible in the land of interest group politics and the capitalist ethos? Where is the flaw in the left-wing critique? What resources could exist within liberal capitalist culture for taming technology? Some answers emerge on the basis of the particular characterization of modernity and technology developed above.

It is a great, if common, mistake to assume, as the left-wing critique does, that in a pluralist liberal democracy all interest groups must be concerned with immediate, selfish goods. On the contrary, in modern societies, where life is grounded on the idea of progress, a concern for the future forms some part of everyone's thinking, and future-directed, ideological interest groups tend rather to proliferate. Certainly there is no structural reason, then, why an environmental movement could not arise in a liberal society. To see why it became in fact the ideological growth stock of the eighties and beyond, one might look at some further factors.

As we have seen, when technology comes to be viewed as an independent and dangerous force, the will to control it inevitably emerges from out of the technological attitude itself. Technology is, as it were, a self-regulating movement. And the fear of technology, once thus begun, actually fits perfectly into the liberal capitalist world-view, strange as that may seem. Skeptical regarding the good and happiness, liberal societies tend to measure progress negatively, in terms of the triumph over evil—over the barbarism and backwardness of the state of nature. But the "negative orientation" of the liberal mind has one inevitable shortcoming: due to the very progress of society, these natural evils gradually recede, losing their power to ground and orient our lives. New evils must be found to take their place. That is, I believe, the psychological and existential meaning of liberal environmentalism. Ecological disaster has taken the place of the state of nature. The middle classes of the advanced industrialized nations, who can no longer feel the old, bourgeois sense of pride in the triumph of civilization over barbarism (and who have seen the demise of fascism and communism), have redefined progressivism in terms of saving the planet.

Yet it still seems, from the perspective of left-wing common sense, that this should not be possible. Liberal capitalism embraces a world of competition and strife, not of harmony with nature. Wholly committed to growth, technology, and the unbridled exploitation of nature, it contains no countervailing principle that might rein humanity in. It simply lacks the resources to check the human quest for power.[25]

The error here is to assume that liberalism's hostility toward nature entails an unlimited trust in humanity. Liberalism's deepest root is precisely a fear of man: we are not by nature sociable or good for each other; our original condition is a state of war; government has been invented to rein humanity in; and this invention itself must be carefully restrained through popular representation, separation of powers, constitutionally grounded rights, and so forth. In the economic realm, as well, there must be no monopoly, whether public or private, but rather free competition and mutual checking. The core of liberalism, in short, is a distrust of all unchecked human power. (This in contrast to the socialist tradition, with its greater trust in the goodness of man and of human power, so long as the latter is centralized and public). Thus, it is not unintelligible or even surprising that our traditional ethos of "limited government" should have expanded in recent decades to include the need for "limited technology."[26] As shocking as it may be to long-accepted ideas, the liberal capitalist psyche turns out to be uniquely fertile soil for the growth of a certain kind of environmentalism—that based on the "negative idealism" of triumphing over disaster, on the deep suspicion of human power, and on the will to control technology resulting dialectically from the technological impulse itself.

The foregoing arguments are not meant to prove, however, that technology necessarily leads to liberal democratic capitalism or that the latter must inevitably succeed in taming the former. They only seek to call into question the long-prevalent theory that these outcomes are essentially impossible. They do further suggest, however, that whether freedom or despotism shall ultimately triumph, or unintended consequences be controlled, will be determined by the fortuities of moral and political life and not by the automatic unfolding of technology's supposed inner logic.

THE PROBLEM OF TECHNOLOGY ON THE RIGHT

There is a further set of problems, however, that seems more strictly tied to the phenomenon and fate of technology as such. Cultural conservatives, rooted in both the religious and the romantic-existentialist Right, have long opposed technology as inevitably leading to aesthetic, moral, and spiritual degradation (themes also picked up, of course, by the Left and especially the New Left).

Even if environmental disaster is avoided and every trace of oppression eliminated, the argument goes, the unhindered success of technology necessarily leads to the expansion of the material world, the deepening spread of consumerism, the endless production of new goods and needs; and all of this distracts and seduces man from his higher vocation, threatening the permanent atrophy of his nobler faculties. Universal affluence and ease will cause men to become spoiled and decadent, robbing humanity of the indispensable lessons of adversity, and fostering a world of lethargic and undisciplined mass men. At the same time, all the beauties of external nature and all the living artifacts of traditional culture will be razed to make room for new factories, condos, and shopping malls—a sterile landscape to accompany an empty heart.[27]

Still more deadly to the soul than these material effects of technology, however, is the thought or attitude that lies behind it. Amidst even the greatest natural splendors, the mind feels little if it sees only the inanimate world of modern physics. What true belief can there be in romantic love or political greatness on the basis of our reductionist psychology; what notion of human freedom and dignity given our determinist natural and social science? No lofty human possibility can flourish in the disenchanted, leveled, homogeneous, quantitative world of modern scientific rationalism. And yet, the argument continues, this life-destroying rationalism is not imposed on us by the nature of reality itself but is rather something we impose on ourselves through our will to technological mastery. We ourselves will this nihilistic world without quality, form, or hierarchy—for such the world must be if we are to place it entirely at our disposal. Thus, in the end, it is the technological attitude that has stifled our souls by expelling from the world every possible experience of beauty, wonder, and reverence.

These two general views may be said to represent, respectively, the materialist and idealist accounts of the emergence of the soulless "mass man." Since the rise of European fascism, political theorists have not pushed such arguments with the old passion (except on the Left), but there is still a good deal of common sense and popular opinion on this side. Certainly among third world and other external observers, it is almost axiomatic that wealth and materialism have made the West spoiled, soft, shallow, and decadent. And we ourselves seem fairly obsessed by the thought that our brave new world is creating a rebarbarized population of yuppies, couch potatoes, and valley girls.

Yet the question that needs to be raised here is, as above, not whether these evils are a danger, but only whether they follow inevitably from technological progress, so that the "problem of technology" may be declared the crucial, underlying political and philosophical issue.

As to the first, "materialistic" theory of mass man, most would concede that—like the left-wing critique—it has not, as yet, been borne out by the facts. Despite several generations now of historic levels of affluence, the Western democracies continue to show some strength of character and have thus far been able to meet the challenges of all those presuming upon their irreversible decadence.[28] One possible explanation for this, which meets the materialist argument on its own terms, is that the rapid advance of technology brings in its train a great deal of social and economic disruption, so that it inevitably produces almost as much "adversity" as it removes. Thus, if modern man has not yet sunk into the bovine happiness of the herd—disappointing long-standing expectations—that may be because, with the concomitant erosion of neighborhoods and families, the rise of crime and divorce, and the constant threat of economic obsolescence, he does not yet feel terribly spoiled. One might also argue, on a somewhat brighter note, that capitalism, with its much-decried tendency to engender competition, has for this reason proven remarkably resourceful in spurring energetic activity and accomplishment (not to say workaholism) in the very midst of prosperity. At least so far.

Looking to the longer term, one could of course point to the fact that, in the past, leisure classes have existed that were not corrupted, but rather liberated and ennobled, by great affluence. To be sure, this required the existence of a strenuous, uniform, socially enforced, and usually aristocratic moral code, which is unlikely to return in our liberal, pluralistic, egalitarian, materialist culture. But the laws of the human spirit are still so unknown that no one can say with confidence that some new basis—perhaps religious or mystical—cannot emerge for the ennoblement of leisure.[29] In every age, moreover, there have always been at least isolated individuals who, owing perhaps to a certain innate strength of mind or character, were able to rise above their circumstances and answer the call of their highest possibilities. And it is not unlikely that technological progress will someday give us the power—through drugs or genetic engineering—to produce that innate strength and so to liberate humanity from the influence of the otherwise stunting cultural conditions that such material progress also brings.[30]

As for the second account of mass man, which puts the blame on scientific rationalism stemming from technology, could one not respond with a point similar to that made above regarding liberal environmentalism: this romantic-existentialist critique of technology is in fact a dialectical expression of the technological attitude itself. Consider its origins.

Enlightenment rationalism and modern science, which arose with the hope of controlling nature for the sake of man's physical well-being, had the "unintended consequence" of undermining his psychic or emotional well-being. He felt homeless and estranged, his spirit could not breathe in the depleted metaphysical environment created by modern physics. But to this predicament, a number of responses would have been possible. One could have accepted this bleak world with resignation and with the renunciation of human mastery—either on the model of Pascal, for example, or of Lucretius, who found a new, austere happiness by adjusting his soul to the harsh truths of ancient atomism.

Instead, however, the problems of modern science were addressed in the modern spirit of indignation and rebellion, as if humanity had a right to a cosmos that supported its noblest passions. Our whole understanding of the "problem of technology" was itself formulated from the standpoint of man's quest to master all the conditions of his existence. And consequently its solution was sought through a further recourse to technology. In Romantic thought, the artistic side of human nature was elevated above the rational: man's essence was relocated in "creativity." It is within man's power and right, it was argued, to engineer a new interpretation of nature and reason that would answer to the needs of the heart just as the old paradigm had to those of the body. Thus, when the Romantics objected to the scientific-utilitarian exploitation of nature, ultimately that was only because this spoiled it for their own use—as a vehicle for aesthetic experience, a neutral "other" on which to project and unfold the inner self. The scientist sees nature as matter, the Romantic as subject-matter. Both equally seek control, the one over the physical world of objects, the other over the subjective world of perceptions and feelings. Similarly, Nietzsche and the existentialists assert man's capacity to create himself, to legislate values, and to interpret the world in the service of human needs and aspirations.[31] In short, the problem of science and of the mass-man is solved by extending technological control into the realm of what we now call "belief-systems."[32]

If this solution is correct, however, then the problem of mass-man derives not from technology as such but only from its incomplete application. Technology comes to light once again as a self-correcting phenomenon—hence not as the central political and philosophical problem.

On the other hand, one may question the adequacy of this solution. If we follow it and free our minds from the seemingly stifling constraints of scientific rationalism, perhaps a more elevated and authentic culture will come to flower; but perhaps mass culture will only triumph all the more fully. The tribunal of reason may hold us back, but it also holds us up. And the mass-man, who chafes under every form of authority and discipline, may celebrate the demise of "Truth" by sinking still farther into bovine self-satisfaction. Liberated from bondage to reason, he is now free to believe whatever he wants—whatever makes him feel good about himself—because, as the ultimate expression of modern technology, he has placed his own beliefs at his disposal.

I am not arguing, of course, that either of these two scenarios is inevitable, but only that both are possible. Once more, it seems that the inner logic and future consequences of technology are far more indeterminate than its critics tend to suppose, and that independent moral and intellectual forces will continue to play a crucial role in shaping human destiny.

Conclusion

From this overview of the characteristics and problems of modern technology one cannot formulate firm conclusions; but it will not be inappropriate to suggest some general lessons and directions for thought.

We have seen that modern technology is indeed highly problematic; but it poses many different problems, some of them mutually exclusive, and it seems impossible to say which will loom largest in the future. Paradoxically, the more one surveys the large and fairly booming field of technology criticism, the more one is driven to conclude that the meaning and destiny of technology is indeterminate. And if this conclusion is correct, then there simply is no single issue that can be identified—and placed ceremoniously at the center of our thought—as the essential "problem of technology."

This is not to deny that technology is probably the most characteristic and imposing feature of our age; but just for that reason we are probably too inclined to attribute to it a specific, coherent, and all-determining significance. We should be more on our guard. If it is the true task of philosophy to liberate itself from the prejudices of the age, then perhaps our specific duty is to free ourselves from technology—not only from the deification but also the demonization of it (the true *idée fixe* of our time).[33]

In addition to being theoretically questionable, moreover, this demonization—the excessive emphasis on the "problem of technology"—brings a specific problem of its own. When technology is viewed as the central and defining philosophical issue, elementary moral and political distinctions tend to lose their clarity and force. Liberal democracy, communism, and fascism, for example, appear fundamentally equivalent, since they are equally technological. Furthermore, to the extent that we view ourselves as helpless pawns of an overarching and immovable force, we may renounce the moral and political responsibility that, in fact, is crucial for the good exercise of what power over technology we do possess. Thus, an exaggerated philosophical concern with "the problem of technology" hinders the effective management of the various real problems of technology.

At or near the heart of the modern technological turn, I have suggested, are two things: the idea of progress and the radical subordination of ends to means resulting from the dismissal of the idea of a *summum bonum*. We might begin to distance ourselves from technology, then, by questioning these two things—first, by attempting to overcome our "forgetfulness" of happiness and the human good; and second, by adopting—as part of this very attempt—a more humble attitude toward the past, especially the pretechnological past, both Eastern and Western. For, in such earlier thinkers as, say, Confucius and Aristotle one finds a comprehensive theory of happiness formulated in serene detachment not only from the turbulent hopes of modern technology but, just as importantly, from the angry rebellion against it. With their help, we might perhaps begin to free our minds from Western modernity and, at the same moment, from modern self-loathing, of which "the problem of technology" is the greatest manifestation.

The same conclusion also emerges as follows. It is often said that early modern philosophy was "polemical" in that it arose, not from a direct encounter with life, but from a reaction against existing ideas—against "the kingdom of darkness from vain philosophy [i.e., Aristotle] and fabulous traditions [Christianity]" (in Hobbes's famous phrase). This reaction, moreover, concerned not only the falseness of the prevailing ideology but especially its practical consequences, its use as a tool of exploitation. Most of later modern thought has remained polemical—and indeed technological—in this sense: it grows, not from a naive pursuit of truth, but from our famous posture of alienation—a rebellion against the consequences of prevailing ideas, and an activist determination to control and overcome them. Thus we speak of the "counter-Enlightenment," the "Romantic reaction," and so forth. And certainly most of contemporary philosophy—especially all variants of postmodernism—self-consciously derive from a political, moral, or aesthetic reaction against the consequences of instrumental rationalism—against "the problem of technology." But the more we rail against technology, the more we remain in its grip.

What seems to be needed is rather a nonpolemical form of thinking that takes its start, not from a settled opinion regarding the danger of prevailing beliefs, but from a more naive, honest, and direct encounter with life—an encounter that, following the model of an earlier, pretechnological mode of thought, draws upon the elemental and permanent human concern for happiness. Doubtless, there are problems with technology, but we will never adequately comprehend or respond to them if we do not begin by looking away from them to questions that are currently less urgent and imposing but are perhaps more honest and fundamental.

Notes

1. Martin Heidegger, "The Question Concerning Technology," in *The Question Concerning Technology and Other Essays,* trans. William Lovitt (New York: Harper and Row, 1977) and *Nietzsche,* vol. 4: "Nihilism," trans. David Krell (San Francisco: Harper and Row, 1987).

2. Consider, for instance, Friedrich Engels, *Socialism: Utopian and Scientific,* trans. Edward Aveling (New York: International Publishers, 1935).

3. See, for example, Max Horkheimer, *Eclipse of Reason* (New York: Seabury, 1974) and Herbert Marcuse, *One-Dimensional Man* (Boston: Beacon, 1964).

4. For a very useful survey of views on the phenomenon and problem of technology from a different point of view see Carl Mitcham and Robert Mackey, *Philosophy and Technology: Readings in the Philosophical Problems of Technology* (New York: the Free Press, 1972), 1–30.

5. Hannah Arendt puts particular emphasis on this factor since it forces the worker for the first time to submit to the alien motion and rhythm of the machine. See *The Human Condition* (Chicago: University of Chicago Press, 1958), 144–53.

6. On this last point, see especially Albert Borgmann, *Technology and the Character of Contemporary Life: A Philosophical Inquiry* (Chicago: University of Chicago Press, 1984).

7. See Jose Ortega y Gasset, "Man the Technician," in *Toward a Philosophy of History* (New York: W. W. Norton, 1941) and *Revolt of the Masses,* trans. Anthony Kerrigan (Notre Dame: University of Notre Dame Press, 1985).

8. On this theme see the beautiful discussion by Hans Jonas in *Philosophical Essays* (New York: Prentice-Hall, 1974), 3–20. See Also Andrew Goudie, *The Human Impact on the Natural Environment* (Oxford: Basil Blackwell, 1981), Carolyn Merchant, *The Death of Nature: Women, Ecology, and the Scientific Revolution* (New York: Harper and Row, 1980), and William McKibben, *The End of Nature* (New York: Random House, 1989). Although long oblivious of these threats, today we often seem to be in danger of overestimating the fragility of the earth.

9. See Jacques Ellul, "The Technological Order," in *The Technological Order,* ed. Carl Stover (Detroit: Wayne State University Press, 1963) and *The Technological Society,* trans. John Wilkinson (New York: Knopf, 1964); see also Langdon Winner, *The Whale and the Reactor: A Search for Limits in the Age of High Technology* (Chicago: University of Chicago Press, 1986), 47–54.

10. Maximilien Robespierre, "Report on Religious and Moral Ideas and on the National Festivals," in *Oeuvres Complètes,* 10 vols.(Paris: E. Leroux, 1910–67) vol. X, 444.

11. Michael G. Kammen, *A Machine That Would Go of Itself: The Constitution in American Culture* (New York: Knopf, 1986).

12. See Bacon, *The Great Instauration*, vol. 4, 24, and *The Wisdom of the Ancients*, vol. 6, 721, in *The Works of Francis Bacon*, ed. Spedding, Ellis, and Heath, 14 vols. (London: Longman etc., 1857–74); Thomas Hobbes, *Leviathan*, ed. Michael Oakeshott (Oxford: Basil Blackwell, 1957), 435; Descartes, *Oeuvres et Lettres*, ed. A. Bridoux (Paris: Pleiade, 1952), 168.

13. See Hans Jonas, *The Phenomenon of Life: Toward a Philosophical Biology* (New York: Dell, 1966), 92–95, 200–207; and Alexandre Koyré, *From the Closed World to the Infinite Universe* (Baltimore: Johns Hopkins University Press, 1957). Consider also Heidegger "Science and Reflection," in *The Question Concerning Technology*. For the opposite view see Stephen Toulmin and June Goodfield, *The Architecture of Matter* (Chicago: University of Chicago Press, 1962), 39: "If, during the last few decades, scientists have at last been able to supplement, and even to revolutionize, the age-old craft traditions, this has been an incidental—and largely uncovenanted—fruit of scientific success."

14. On these general themes, see Mircea Eliade, *Cosmos and History: The Myth of the Eternal Return*, trans. Willard Trask (New York: Harper and Row, 1959); *Myth and Reality*, trans. Willard Trask (New York: Harper and Row, 1963); Leo Strauss, "Progress and Return," in *The Rebirth of Classical Rationalism: An Introduction to the Thought of Leo Strauss*, ed. Thomas Pangle (Chicago: University of Chicago Press, 1989); and Edward Shils, *Tradition* (Chicago: University of Chicago Press, 1981). See also Daniel Lerner, *The Passing of Traditional Society: Modernizing the Middle East* (Glencoe, Ill.: The Free Press, 1958); and Kranzberg and Pursell, eds., *Technology in Western Civilization* (New York: Oxford University Press, 1967), vol. I, part I.

15. On the subject of ancient "primitivism," see Arthur Lovejoy and George Boas, *Primitivism and Related Ideas in Antiquity* (Baltimore: Johns Hopkins University Press, 1935). See also Arthur Melzer, *The Natural Goodness of Man: On the System of Rousseau's Thought* (Chicago: University of Chicago Press: 1990), 55–57.

16. Plutarch, "Life of Marcellus," in *The Lives of the Noble Grecians and Romans*, trans. John Dryden and rev. Arthur Clough (New York: Modern Library, 1932), 378.

17. See James H. Nichols Jr., *Epicurean Political Philosophy: The De rerum natura of Lucretius* (Ithaca: Cornell University Press, 1972), 101–78.

18. Thomas Hobbes, *Leviathan or the Matter, Form, and Power of a Commonwealth, Ecclesiastical and Civil,* ed. Michael Oakeshott (Oxford: Basil Blackwell, 1957), 80.

19. Ibid.

20. See Langdon Winner, *Autonomous Technology: Technics-out-of-Control as a Theme in Political Thought* (Cambridge: MIT Press, 1977) for a good discussion.

21. The distinction between "Left" and "Right," never very clear, becomes less so with each passing year. It is employed here because it remains more useful than any equally simple distinction that could be put in its place.

22. Jonas, *Philosophical Essays,* 18–19. Robert Paul Wolfe, "Beyond Tolerance," in *A Critique of Pure Tolerance* (Boston: Beacon, 1970), 49–52.

23. See especially Erich Fromm, *The Revolution of Hope: Toward a Humanized Technology* (New York: Harper and Row, 1968); Lewis Mumford, *The Myth of the Machine,* 2 vols. (New York: Harcourt, Brace, Jovanovich, 1967–70); Marcuse, *One-Dimensional Man* (Boston: Beacon, 1964); Theodore Roszak, *The Making of a Counter Culture: Reflections on the Technocratic Society and its Youthful Opposition* (Garden City, N.Y.: Doubleday, 1969); Horkheimer, *Eclipse of Reason* (New York: Oxford University Press, 1947).

24. On this practical level, one must also question the too easy equation of technology and science. Is there not truth to the older, Enlightenment view that, whatever effect technology may have at the moment, the underlying scientific enterprise requires and fosters certain moral habits that are favorable to individual liberty and democracy, such as skepticism, dissent, tolerance, creativity, open debate, objectivity, and love of truth? See the classic statement by Jacob Bronowski, *Science and Human Values* (New York: Harper and Row, 1965). See also Paul Goldstene, *The Collapse of Liberal Empire: Science and Revolution in the Twentieth Century* (New Haven: Yale University Press, 1977), 91–127; and Yaron Ezrahi, *The Descent of Icarus: Science and the Transformation of Contemporary Democracy* (Cambridge: Harvard University Press, 1990).

25. See Roszak, *Making of a Counter Culture* and Merchant, *The Death of Nature.*

26. This argument, of course, makes no sense on the left-wing interpretation of liberalism, according to which the principle of limited government was never anything but a secret means for removing limits from economic power. It is true that liberalism prefers to regulate economic activity through market rather than governmental mechanisms and that it regards economic power as less dangerous than political, but that does not mean that liberals are insincere in their distrust of human power as such.

27. See, for example, Ortega y Gasset, *Revolt of the Masses;* Henri Bergson, *The Two Sources of Morality and Religion,* trans. R. Ashley Audra and Cloudesley Brereton (New York: Henry Holt, 1935), 298, 302–309. See also Borgmann, *Technology and the Character of Contemporary Life;* Alexis de Tocqueville, *Democracy In America,* trans. J. P. Mayer (New York: Doubleday, 1966), Pt. II, Bk. 2, chaps. 10, 11, 16, 17; Bk. 3, chaps. 17, 19, Bk. 4, chap. 6.

28. In *The Present Age,* Kierkegaard writes that his generation has lapsed "into complete indolence. Its condition is that of a man who has only fallen asleep towards morning: first of all come great dreams, then a feeling of laziness, and finally a witty or clever excuse for remaining in bed.... In the present age a rebellion is, of all things, the most unthinkable. Such an expression of strength would seem ridiculous to the calculating intelligence of our times..." trans. Alexander Dru. (New York: Harper and Row, 1962), 35. This was written two years before 1848. See also Solzhenitsyn's predictions for the outcome of the cold war in his "Harvard speech," *A World Split Apart* (New York: Harper, 1978).

29. See Bergson who argues precisely that technology will lead to mysticism (*Two Sources of Morality and Religion,* 300–303, 308–17).

30. "The mind depends so much on the temperament and on the disposition of the organs of the body that if it is possible to find some means which generally renders men wiser and more skillful than they have been hitherto, I believe that it is in medicine that one must search" (Descartes, *Oeuvres,* 168–69).

31. This, of course, is where Heidegger turns away from and against Nietzsche and existentialism. Without trying to judge whether Heidegger himself is successful in his attempt to free himself from technology, I would suggest that most of those today who attack technology in his name in fact remain closer to Nietzsche.

32. See Carl Schmitt's *Political Romanticism,* trans. Guy Oakes (Cambridge: MIT Press, 1986). For a discussion of related ideas see Jeffrey Herf, *Reactionary Modernism: Technology, Culture, and Politics in Weimar and the Third Reich* (Cambridge: Cambridge University Press, 1984), 1–17, 109–29.

33. It will of course be objected that it is impossible to escape the ideas of one's time; but this familiar thesis has never been proved conclusively and probably cannot be. Thus, given our necessary uncertainty on this score, it seems best to act as if we were capable of freedom and see what develops, since to assume the opposite is to make the thesis self-fulfilling through a preemptive surrender to prevailing beliefs.

8

Global Technology and the Promise of Control

Trish Glazebrook

Technology theorists are remarkably silent on the topic of globalization. Although philosophy of technology is burgeoning as a discipline, its proponents have little to say about technology transfer to developing nations, and the impact on the global human condition of technology outside the West, or, as it is also called, the North. There are exceptions, of course, most notably Sandra Harding, whose work on post-Enlightenment, postcolonial science and technology is extensive.[1] Likewise, much has been done in social geography focusing primarily on communications technology. The lack of assessment of the global implications of technology in science studies, cultural studies, and technology studies gives pause for thought. Are contemporary philosophers of technology simply reproducing the ethnocentrism evident in science and technology themselves? Or does the globalization of technology call for new ways of thinking about traditional philosophical questions, for example, the political consequences of the metaphysics and epistemology that underwrite modern technology, that philosophers are only now beginning to envision? I will argue that the answer to the second question is yes, and I will show that the task such innovative thinking entails presses urgently if the answer to the first question is not to remain, yes. For without careful reflection on the ideology that is and informs modern technology, the technological dream of controlling nature spills over into political and social practices of exploitation.

I make my argument in four parts. First, I show how technology has figured in traditional philosophical treatments of the question of what it means to be human. Second, I argue that the distinction between science and technology, so clear for Aristotle, is blurred in modernity, and that the modern techno-scientific project is underwritten by a logic of domination and control. Thirdly, I situate technology in ethical, political, and cross-cultural practices. I show how forestation programs in India were disastrously destructive because of their failure to respect local knowledge, and I use the Bantu education policy in South Africa to demonstrate that the introduction of science and technology to non-Western cultures has been and potentially remains a process not of democratic access, but of cultural subjugation. Finally, I look at the collapse of the dream of control in two contexts. The nuclear power industry in the West shows that the cooperation of national and techno-scientific, corporate interests is a dangerous complicity that promotes an illusion of control. Yet the French experience with videotex stands as an example of the human ability to transcend federally directed initiatives and to appropriate technology democratically, as the experience of AIDS patients demonstrates the ability of an informed public to intervene in the institutions of technology. In conclusion, I suggest that the globalization of technology can serve to perpetuate ideological and practical control of both nature and human beings, but offer the optimistic possibility that this need not be a foregone conclusion.

Global technology is not novel but rather is an entrenched practice, and the ethical and political task at hand in the present is to integrate local concerns within globalized culture. Global technology has a substantial role to play in determining the human experience, and therefore its ethical, social and political implications demand ongoing philosophical analysis and dialogue, lest a definitive function of human being remain a global mechanism for reinforcing privilege and hegemony over the values of community and democratic empowerment.

Knowledge as the Human Project

Aristotle begins his *Metaphysics* with the claim that "[a]ll humans by nature desire to know."[2] Likewise, when Descartes asks, "What am I?" in the second of his *Meditations,* he concludes, "A thing that thinks."[3] The capacity for knowledge, our ability to think, has been definitive of human being in the tradition of Western philosophy since its very beginning, and remains so in modernity. If we are what we eat, then human being continues to demonstrate an inexhaustible appetite for knowledge. For many in the West/North, our day has the consumption of information built right into it. We read the paper, watch the Nature or History Channel, and spend both work and leisure time cruising the information superhighway on the Net. That such knowledge is not just a human function among others, or a specialty of us moderns, is evident in the Christian ideology of sexuality: Eden was lost through carnal knowledge, and we still talk of knowing in the biblical sense. Nor are we Westerners/ Northerners alone in constructing ourselves socially as knowers. Around the world, peoples define themselves on the basis of their cultural storehouse of knowledge, and build their practices and calendars on the basis of preserving and disseminating knowledge through ritual.

What is knowledge, however? This is a broad question, and even if it could be articulated thoroughly, no final answer would then have been given, since knowledge is a process that evolves with culture and history. My task is precisely to trace the evolution of technology as an ideology and practice of knowledge in order to place into relief a tension between Western/Northern technology and global interests of politics and ethics. Hence, I confine myself to an analysis of the place of technology in knowledge, and the history I tell begins with Aristotle. He was quite clear about what knowledge is, and drew the original distinction between science and technology that is fundamental to Western intellectual history. In the *Topics* and the *Metaphysics,* he divides knowledge into three kinds: theoretical, practical, and technical.[4] He differentiates them on the basis that each has a different end. The end of theory is simply knowledge itself. This is what one knows just for the sake of knowing it, and herein Aristotle includes metaphysics, mathematics, and natural science.[5] The end of praxis, which includes ethics and politics, is action. And the end of technical knowledge, production, is the thing that is made. Carpentry, for example, has as its end the house that is built, and likewise medicine is directed at and for the sake of health. Aristotle's word for "nature" is *physis* and his word for "production" is *technê.* Hence, the etymological origins of physics and technology are evident in Aristotle's taxonomy. In the next section, I will argue that the Aristotelian difference between theoretical science and technology is blurred in modernity, but first I make clear here the implications of his distinction.

The first thing that must be noted is that technê is an ambiguous term. It means both the things that are made in production, and the knowledge by which they are made. Hence, it can have ontological or epistemological force. I am very much persuaded by Heidegger's insight that technology is not just a collection of equipment, but "a way of revealing."[6] In fact, this is the basis for my analysis that technology is and can remain a global mechanism for reinforcing privilege and hegemony over democratic empowerment. Hence, when I speak of technology, I do not mean particular items of technological equipment, but rather the ideology that permeates the production and use of such equipment.

Nonetheless, I work within Aristotle's ambiguity in that I begin with a question of ontology: how are artifacts different from natural things? A debate is currently underway in philosophy concerning this question,[7] but it is centered on restoration ethics[8] rather than Aristotle's ontological distinction. Aristotle argues at *Physics* 2.1 that what is definitive of natural things is that "they have within themselves a principle of movement (or change) and rest."[9] He goes on at 2.8 that nature is therefore teleological in that, rather than natural things developing by chance, they grow toward some end. Together these claims say that, for example, an acorn moves itself toward its final cause, which is an oak tree. Artifacts, however, have no such internal principle of growth. Production begins with the artist's conception of what is to be made.[10] Aristotle draws the odd conclusion that an artist chooses material "with a view to the function, whereas in the products of nature the matter is there all along."[11] What he means is that the relation between form and matter is necessary in the case of nature (wood does not exist except in trees; trees cannot but be made of wood), but not so in the case of an artifact (gold can be made into jewelry or a statue; a statue can be made of gold or bronze). Natural form is a principle of self-directed material development, while artifacts have a form that is imposed on matter by the artist.

The point to draw from Aristotle's analysis is that technology depends on nature in a way that nature does not depend on technology. For an artifact must always be made from some material appropriated from nature. Things taken from nature by the artist and formed into an artifact are at most an interruption of natural process, which persists. Wood, for example, rots despite its treatment by the artist. Thus, technology is a derivative way of being, and control of natural processes is at best partial and temporary. Yet modern technology is an ideology of control and manipulation that has lost sight of this Aristotelian truism. A succinct example of this modern oversight presented itself when I recently read a paper called "Nature versus Technology" precisely to make the Aristotelian point. Someone joked, "I've got my money on technology!" The point I made, of course, was that in fact technology is never removed from but rather always deeply embedded in nature, and that therefore, it is nature that will always have the last word, so to speak.

Science, Technology, and the End of Nature

Two pivotal conceptions of nature since Aristotle figure centrally in the illusion that technology promises control over nature. The first is Christian, the second scientific. Both these moments are complicit in a blurring of the distinction between nature and artifact such that human being comes to understand itself to be able to overcome and dominate nature. Nature is of course conceived as an artifact in a quite straightforward way in Christian theology. It is the product of a divine artisan. More significantly, however, the story of the expulsion from the Garden of Eden at Genesis 3:17–24 establishes the divine mandate of earthly dominion for human being. Adam is sent from the garden with the instruction to till the ground and toil for sustenance. He has become like a god in eating of the Tree of Knowledge, and is subsequently empowered to manipulate his environment to meet his own needs. Modern science has roots that are deeply embedded in Christian theology, as indeed the Enlightenment project arose from this religious context.[12] What I show in this section is that Bacon's account of technological science endorses a logic of domination and control. The Newtonian substitution of a mechanistic worldview in place of Aristotle's teleological conception is complicit with the Judeo-Christian directive and the Baconian program. Once nature is freed of teleology, it is readily conceived as serving no end other than human. Hence, the ideology of modern science lays nature bare for its technological reduction to resource.

Francis Bacon published his *Great Instauration* in 1620. His plan for science was to "conquer nature in action,"[13] and he established his experimental task as inquisition.[14] He claimed to "be content to wait upon nature instead of vainly affecting to overrule her,"[15] yet his explicit goal was "that the mind may exercise over the nature of things the authority which properly belongs to it,"[16] by extracting conclusions "out of the very bowels of nature"[17] that is "under constraint and vexed; that is to say, when by art and the hand of man she is forced out of her natural state, and squeezed and moulded."[18] Bacon founded the modern experimental method in metaphors of domination and torture. His logic of scientific methodology did not allow nature to speak freely, as it were, but subjugated it to his rule. Baconian science is thus not knowledge for the sake of knowledge, as Aristotelian science was. Rather, it is the groundwork for improving the human condition by manipulating nature technologically.

Newton's work, *Mathematical Principles of Natural Philosophy*, first published in 1684, is substantially different in tone. He uses no metaphors of warfare and torture. Yet his methodology remains experimental.

He is interested in the manipulation of nature under idealized conditions determined and constructed by the scientist, rather than the observation of entities in their natural context that pervades, for example, Aristotle's biology. But of course, physics has by Newton's day changed its object. Whereas Aristotle was concerned with *ta physika,* things in nature that move themselves, such that biology constituted a substantial part of his inquiry into nature, Newton's laws govern bodies subject to forces. Newtonian science is paradigmatically physics, concerning which his hypothetico-deductive experimental method seeks eternal and unchanging truths. His "mathematical principles of philosophy"[19] homogenize natural entities into bodies whose properties "are to be esteemed the universal qualities of all bodies whatsoever."[20] Rather than letting nature inform his epistemology, he wishes to "derive the rest of the phenomena of nature"[21] from his universal, rational principles. Accordingly, Newton's text sets a standard for objectivity through mathematical idealization that distinguishes bodies not according to their teleology, that is, different qualities and functions, but only by quantifying them in terms of their universal attributes such as extension, hardness, impenetrability, and gravitation. Physics has thus come a long way from nature. In fact, the relation between particle physics and the everyday objects of the human lifeworld is a problem even for physicists.

Furthermore, Newton's mechanistic worldview in which nature is conceived artifactually as a giant clockwork, rests on severely narrowed conceptions of motion and causality. Whereas Aristotle thought about nature first and foremost in terms of final causes, Newton's concern is limited to moving causes. Motion no longer includes growth, as it did for Aristotle, but is limited to locomotion. Newton wants to know how external forces govern rest and change of place, not how internal drives govern growth, and his laws of motion clearly govern locomotion. His worldview is mechanistic and his method is hypothetico-deductive. This conjunction of metaphysics and epistemology led Goethe to engage in a lifelong polemic against Newtonian science precisely because it envisioned nature in a reductive way.[22] Newton affirmed ideologically in his scientific practice that nature consists in nothing more than bodies subject to forces that can be controlled and manipulated in experimentation. Technological mastery of natural entities is the logical conclusion of a science that has no place for natural teleology. The objectification of nature renders it already available conceptually to be nothing more than manipulable resource. Thus, modern science in its incipience promises and claims to deliver human mastery over nature. No longer content simply to know nature, theoretical science is inherently technological in its usefulness for application. It is techno-science.

Where does Aristotle's third division of knowledge stand in this techno-scientific constellation? The mechanistic model separates ethical and political questions from scientific. The historical rise of science as technology determines the very essence of human being as knowers in modernity, and it displaces ethical questions and insights out of the heart of the human project of knowing. Ethical wisdom is neither quantifiable nor objective. Yet there are thinkers who argue that knowledge cannot be evaluated simply in terms of its truth or falsity. Maria Lugones and Elizabeth Spelman, for example, argue that "theories appear to be the kinds of things that are true or false; but they are also the kinds of things that can be, e.g., useless, arrogant, disrespectful, ignorant, ethnocentric, imperialistic."[23] In a global context, it is certainly clear that the ideology and practices that constitute technology demand rethinking in ethical and political terms.

Globalized Technology in Its Social and Political Context

That Western techno-science gave rise to colonial practices that were both racist and unsound ecologically is readily evident in the attitude of Robert Boyle, well known for his law relating pressure, temperature, and volume, and less known for his governorship of the New England Company. He explicitly wished to rid the New England natives of their perception of nature "as a kind of goddess," which he saw as "a discouraging impediment to the empire of man over the inferior creatures of God."[24] Devaluing indigenous and local knowledge and practice as standing in the way of progress, as primitive and uninformed, because unscientific, has been a global colonial tradition. What I wish to show in this section is that the globalization of technology can and has been an unethical process. And further, that its unethical shortcomings are not simply the result of naïve faith in Baconian science, as we might perhaps understand if not justify Boyle's view, but rather that globalized technology is entangled in a web of corporate interests and government policies in which vested interests have precluded democratic distribution of the promised benefits. In short, technology has not been globally implemented for the benefit of humanity, so much as for the benefit of an elite. The first part of this section treats eucalyptus planting in India; the second, the Bantu education policy in South Africa.

Vandana Shiva details how eucalyptus reforestation programs in India were based at best on selectively chosen data that do "not reflect the field reality and [do] not satisfy minimum scientific criteria."[25] Ignoring local farmers' claims that the eucalyptus plants were lowering the water table and destroying soil conditions, the president of the Forest

Research Institute of India argued that there is "no scientific basis in the popular fallacy that eucalyptus lowers the ground-water table."[26] In response to challenges to the government policy of emphasis on eucalyptus in 1981, the Forest Department conducted a single-plant experiment on one-year-old juveniles and then implemented a large-scale forestation program, with no regard to the variety of conditions in which the plants would be used and the well-established fact that eucalyptus has different water needs, growth rates, and biomass production at different stages in the life cycle. At worst, then, this program was based on misinformation, and the suppression and falsification of facts. Superficially, it may not seem to be a case of technology. Yet if technology is understood, as I have argued above, as a way of understanding and subsequently manipulating nature toward human ends, then forestry and agriculture are deeply, inherently, and incontrovertibly technological practices.

Furthermore, since, in this case, it was the government of India that established and implemented the eucalyptus policy, it may seem that the issues at stake here are national, not global. Shiva argues, however, that the complicity of the Indian forestry experts in wide-scale desertification served the interests of the pulp industry: "[S]cience and technology have become cognitively inseparable and the amalgam has been incorporated into the economic system."[27] This system is, of course, global capitalism, the economic foundation of Western hegemony, which Shiva argues is "based on exploitation, profit maximization and capital accumulation."[28] Accordingly, the issues of globalized technology, when situated within the context of a developing nation, are not thereby nationalized or localized. Rather, the case of eucalyptus in India demonstrates that globalized technology is embedded in Western hegemonic economic practices. Technology remains an integral part of the Western project of modernity, not despite but because of the Western values its dissemination promotes. This reality may be the greatest threat and weakness of the globalization of technology: while founded on the rhetoric of progress and improvement of the human condition, it can actually serve to further the interests of the few over the many on the basis of Western values of individualism over community, consumption over sustainability, and exploitation over liberation.

This was certainly the case with respect to the Bantu education policy in South Africa. Implemented in 1953, this program represented the culmination of a long history of racist education. Traditional educational practices were efficiently displaced by colonial schools. Opened in the Cape in 1658, the first Western school educated young slaves: they "were to be taught the Dutch language and rudiments of the Christian religion and were encouraged to be diligent, with rewards of brandy and tobacco. They were to be efficient and pliant servants for their new masters."[29] This early moment in colonial educational strategies in South Africa remained the underlying theme, and is significant here in two respects.

First, from its very beginning, their education was never intended to give blacks access to the empowering features of Western technology. Rather, it was and remained directed at sustaining the racist status quo and white privilege by admitting blacks into the techno-scientific super-structure in order that they have the rudimentary skills required to contribute labor to industry. In 1855, Governor Grey argued that education "should try to make them . . . useful servants, consumers of our goods, contributors to our revenue,"[30] and likewise Theal recommended to the Education Commission of the Cape of Good Hope in 1892 that "I would teach the natives to dig the ground and make their furrows straight; also the rudiments of carpentry. They make very good mechanics. . . . If the natives are to . . . be taught at all, they should be taught industry."[31]

Nonetheless, the Apprenticeship Act of 1922 privileged whites in industry by requiring apprentices to have a minimum of eight years of education, that is, to complete Standard 6, which is grade 8 in a 13-grade system. Even by 1970, less than 10 percent of African pupils were achieving that level of education.[32] In fact, half of the six million African children who started school between 1955 and 1968 dropped out before they became literate (Standard 3).[33] The Bantu Education Act was not about enabling Africans to participate fully in Western techno-science, but about maintaining their function as the semiskilled laborers the national economy required. It was not to produce scientists and engineers, but "to train Africans to help man the local administration at the lowest ranks and to staff the private capitalist firms owned by Europeans."[34] If an African wished to study engineering, for example, and thus participate in technology above the level of laborer, he or she had to apply for a permit to attend a white university where such things were offered. In 1960, the Bantu education minister received 190 such applications, of which four were approved.[35] The de Lange Report of 1981 stated that a minority of pupils required preparation for academic study beyond grade 10, whereas "50–80 per cent of children in standards 5–8 [grades 7–10] receiving vocational education . . . is in line with the manpower needs of South Africa,"[36] and therefore recommended dividing high schools into academic versus vocational. Academic education was unfunded, whereas vocational education was funded, and hence African children were relegated to vocationalism and relations of domination and exploitation were maintained. The point of giving access to technology and training in the requisite skills to use it was not to improve the human condition, but to improve the Afrikaner condition by producing a working class, a labor force that could undergird the privileged white class.

Secondly, then, the Bantu education policy made manifest what Heidegger warned of in 1954. He argued that technology is a way of thinking in which "[man] comes to the point where he himself will have to be taken as standing-reserve,"[37] that is, human being itself is reduced to a resource, a means appropriable toward another's end. Hendrik Verwoerd was responsible for implementing the Bantu Education Act. He had claimed in 1948 that the aim of black education should be "to inculcate the white man's view of life."[38] It was not just designed to control and restrict black participation in the social institutions of science and technology, but also to turn blacks into compliant and useful tools in the technological superstructure. Africans themselves were to be indoctrinated into becoming technological entities rather than autonomous users of technology. Education was intended to "passively socialize pupils into the requirements of their future work situation."[39] It was an education for subordination designed to make the underclass content, or at least compliant, with their social role: "[T]he educational system must try to teach people to be properly subordinate and render them sufficiently fragmented in consciousness to preclude their getting together to shape their own material existence."[40] Segregated education was justified because if an African boy got the same education as a white, it made him, "(a) lazy and unfit for manual work; (b) it makes him 'cheeky' and less docile as a servant."[41] In short, it made him into a malfunctional tool.

On the basis of this analysis of India's eucalyptus program and the Bantu education policy, I conclude that the relations of domination that figure prominently in modern technology from its very incipience as techno-science, described in the previous section, have spilled over into political and social relations in the globalization process. The eucalyptus program demonstrates that globalized technology replaces generations of sustainable practice in its complicity with science and antipathy to local, context-sensitive wisdom. The Bantu education policy shows that it has been complicit in imperialism and colonial practices of racism and classism that treat indigenous populations as tools and maintain the institutions that prevent their full participation in the social structures within which they must live. Both histories show that the globalization of technology is deeply entangled in global capitalism, to the detriment of indigenous populations and practices. The technological dream of control is not limited to nature, but has been actualized in cross-cultural interaction and technology transfer such that globalized technology oppresses, controls, and manipulates in social and political contexts. Global technology is the enactment of Western privilege within developing nations insofar as it privileges Western rationality and knowledge over the indigenous, and it sustains Western privilege globally.

Yet the technological promise of control is a false promise. In the next section, I will use the nuclear power industry as an example of the dangerous consequences when the promise of control of nature breaks down. And I will use Andrew Feenberg's analyses of the French experience with videotex and the intervention of AIDS patients in medical experimentation to argue that not only is the promise of social control illusory, in fact users of technology can disrupt social programs by appropriating it toward their own ends. Hence, I conclude my otherwise harsh analysis with an optimism that globalized technology also harbors possibilities for democratization in which technology users can experience it as an empowering and liberating moment of their self-determination.

The Dream of Control

Kristin Shrader-Frechette argues on the basis of her research into the nuclear power industry that control of nature is more limited than the scientific-technological paradigm suggests, in fact even more limited than the scientists involved seem prepared to concede. For the first thirty-five years of nuclear commercial fission, she claims, the problem of the disposal of nuclear waste was not addressed because scientists "were saying that safely isolating the wastes would be easy, once they put their mind to it."[42]

Several Congressional hearings concerning the U.S. Department of Energy's (DOE) nuclear facilities from 1986 to 1989 charged the DOE with a loss of credibility since 1,061 of their sites need massive cleanup costing from $130 to $300 billion. Shrader-Frechette attributes the messes to poor management and storage of waste as a result of DOE managers repeatedly violating their own environmental regulations as well as U.S. law, and claims further that DOE administrators have withheld funds and information from people interested in safety issues, penalized whistleblowers, and failed to spend the money necessary for cleanup.[43] It seems the technologists of nuclear power are not to be trusted.

There is, however, another problem that is epistemological in nature. It has to do with an attitude of overconfidence on the part of scientists and technologists. At Maxey Flats, Kentucky, for example, the corporate leaseholder, Nuclear Engineering Company, who later were renamed US Ecology, claimed in 1963 when their facility opened that it would take plutonium twenty-four thousand years to migrate one-half inch at the site. Plutonium has a half-life of more than twenty-five thousand years. Within ten years, it had migrated two miles off-site.[44] The epistemological problem is that such predictions are based on modeling methodologies that are inadequate to the realities of nuclear waste disposal, and involve inferences that entail subjective judgment for which objectivity is nonetheless claimed. The issues concerning simulation models in hydrogeology include, first, that the data against which models are checked are inadequate and inaccurate since they make predictions, for example, at the Yucca Mountain site, over millions of years on the basis of about a decade of observation. Furthermore, models rely on field conditions that often do not obtain, are typically not testable, and technologists rarely agree on what would confirm them.[45] The technologists allow to pass for scientific finding, argues Shrader-Frechette, what is really scientific opinion. She concludes that the scientific rationality that underwrites nuclear technology is incompatible with an ethical rationality that would include stakeholders' interests and concerns, and responsibility toward future generations.

I argued above that the historical collapse of science and technology squeezed ethics out of the picture, and nuclear waste disposal in the United States is a clear case in point. Technologists have simply overestimated their ability to predict and control natural processes, and have reduced the ethical issues to technical ones. Meanwhile, the failure of the promise of technological control of nature has had and continues to have dramatic ethical consequences. For example, it has been well documented that communities of color and of the poor bear a disproportionate burden of risk with respect to toxic waste.[46] This is also true globally.[47] Technology has a price, which is paid as inequitably as its benefits are shared. The decisions that shape the lives of those who receive an unfairly low benefit, yet pay an unfairly high price, are made in the context of corporate practice and government policy, while affected populations remain uninformed and unconsulted.

Yet if it is clear that control of nature is an overestimated if not false promise, it is equally true that technology users have on occasion shown themselves capable of appropriating technology to their own, more democratic ends. In fact, several theorists have argued that technology can be a source of social empowerment, and that the reality of its authoritarian use and service of elite interests is historically contingent. Andrew Feenberg, for example, has shown how this was the case with the French experience of videotex, an on-line library system that stores pages of information in a host computer to be accessed by users with a terminal and a modem.[48] In the 1980s, more than five million such terminals and modems, Minitels, were delivered to French users. The project was designed to bring France into the information age by means of this hi-tech system of information distribution. Met initially with distrust by a public who saw it as a national service from a highly centralized and controlling state, in the context of a conservative government, it nonetheless became popular quickly. It owed its success to the fact that it was "soon transformed . . . in ways its makers had never imagined."[49] People used it not for information, but as a communication system. In 1982, hackers transformed a technical support facility into a messaging system, and the operators of the service institutionalized the changes. Feenberg argues they were "creating new forms of sociability,"[50] "a new social form"[51] in which people are liberated from social control and instead able to socialize anonymously in private, direct interactions over which they have increased control. Users of the chat space create themselves as virtual subjects who use "that anonymity to shelter and assert their identities."[52]

Similarly, Feenberg argues, AIDS patients were able to intervene in the technocratic organization of medicine through an organized lobby. In the 1980s, the influx of thousands of AIDS patients destabilized the medical system and transformed experimental medicine, previously restricted in its use of human subjects out of paternalistic concern for their well-being, into a standard practice of care for the incurably ill. Rather than "participating in medicine individually as objects of a technical practice . . . more or less compliant to management by physicians,"[53] AIDS patients challenged the system and turned it to new purposes in order "to bring the organization of medicine into compliance with their human needs as participants in the medical world."[54] They intervened and redirected medical technology, and during that process defined themselves as subjects rather than objects in the system, that is, as rational users rather than passive recipients.

If to know is to be human, and technology is the form of knowing that determines human being in modernity, then the recent history of medical technology and of the French experience with Minitel are both examples of democratic self-articulation throughout technological systems in powerful and empowering ways. It seems, then, that the globalization of technology raises questions and offers models for ethically and politically sound cultural practices, despite the ideology of domination and control that marks its incipience and permeates its history. The globalization of technology can serve to perpetuate ideological and practical control of both nature and human beings, but offers the alternative possibility that this threat is not a foregone conclusion. Global technology plays a substantial role in determining the human experience, and its social and political ramifications warrant ongoing and particular analysis, lest this definitive function of human being remain a global mechanism for reinforcing privilege and hegemony over the values of community and democratic empowerment.

Notes

1. Harding has written much herself, and compiled several anthologies: "Feminism, Science, and the Anti-Enlightenment Critiques," in *Feminism/Postmodernism*, ed. Linda Nicholson (New York: Routledge, 1990), 83–106; *The "Racial" Economy of Science* (Bloomington: Indiana University Press, 1993); "After Eurocentrism? Challenges for the Philosophy of Science," in *Proceedings of the Biennial Meetings of the Philosophy of Science Association* 2 (1993), 311–19; "Gender, Development, and Post-Enlightenment Philosophies of Science," *Hypatia* 13: 3 (1998), 146–67; *Is Science Multicultural? Postcolonialisms, Feminisms, and Epistemologies* (Bloomington: Indiana University Press, 1998); Uma Narayan and Sandra Harding, eds. *Decentering the Center: Philosophy for a Multicultural, Postcolonial, and Feminist World* (Bloomington: Indiana University Press, 2000).

2. 980a22. The Loeb translation says, "All men . . ." (trans. Hugh Tredennick [Cambridge: Harvard University Press, 1933]), as does W. D. Ross in his translation in *The Basic Works of Aristotle* (New York: Random House, 1941), but I have chosen to translate *anthrôpoi* as "humans."

3. René Descartes, *Meditations on First Philosophy*, trans. John Cottingham (Cambridge: Cambridge University Press, 1986), 19.

4. 145a14–18 and 157a10–11; 1025b18–26 and 1064a10–19.

5. He distinguishes these again on the basis that they have different objects. The object of metaphysics is being, which is eternal and unchanging; the objects of mathematics, which includes astronomy for Aristotle, are number and relation, which move but are eternal; the objects of natural science are natural things that move in all ways, including generation and destruction. Cf. *Physics*, Books I–IV, trans. P. H. Wicksteed and F. M. Cornford (Cambridge: Harvard University Press, 1929), 198a29–31.

6. Martin Heidegger, "Die Frage nach der Technik," in *Vorträge und Aufsätze* (Stuttgart: Verlag Günther Neske, 1954), 16; "The Question Concerning Technology," in *The Question Concerning Technology and Other Essays*, trans. William Lovitt (New York: Harper and Row, 1977), 12.

7. It began with Robert Elliot's "Faking Nature," *Inquiry* 25 (1982): 81–94, which he has expanded recently in *Faking Nature: The Ethics of Environmental Restoration* (London: Routledge, 1997), but see also Eric Katz, "The Big Lie: Human Restoration of Nature," *Research in Philosophy and Technology* 12 (1992): 231–41, also treated at greater length in his *Nature as Subject: Human Obligation and Natural Community* (Lanham, Md.: Rowman and Littlefield, 1997), and Steven Vogel, "The Nature of Artifacts," manuscript forthcoming.

8. This novel branch of ethics raises questions concerning the technological restoration of nature that has been damaged or destroyed.

9. 192b14.

10. Aristotle, *Nicomachean Ethics*, trans. H. Rackham (Cambridge: Harvard University Press, 1934), 1140a13; *Parts of Animals*, trans. D'Arcy Wentworth Thompson, *The Basic Works of Aristotle*, 640a32.

11. 194b7–8.

12. See David Noble, *A World Without Women: The Christian Clerical Culture of Western Science* (Oxford: Oxford University Press, 1992), although his focus is on the historical genesis of modern science and technology as masculinist projects.

13. Francis Bacon, *The Great Instauration and New Atlantis*, ed. J. Weinberger (Arlington Heights, Ill.: Harlan Davidson, 1980), 16 and 21. This is the introduction to his *Novum Organum*.

14. Bacon, *The Great Instauration*, 2.

15. Ibid.

16. Ibid., 7

17. Ibid., 23.

18. Ibid., 27.

19. H. S. Thayer, ed., *Newton's Philosophy of Nature: Selections from his Writings* (New York: Hafner Press, 1953), 16.

20. Ibid., 3.

21. Ibid., 10.

22. See Dennis L. Sepper, "Goethe Against Newton: Toward Saving the Phenomenon," in *Goethe and the Sciences: A Reappraisal*, ed. Frederick Amrine, Francis Zucker, and Harvey Wheeler (Dordrecht: D. Reidel Publishing Company, 1987), 175–93.

23. Maria C. Lugones and Elizabeth V. Spelman, "Have We Got a Theory for You! Feminist Theory, Cultural Imperialism and the Demand for 'The Woman's Voice,' " in *Women and Values*, ed. Marilyn Pearsall (Belmont, Cal.: Wadsworth Publishing, 1999), 14–24, 20.

24. Qtd. in Brian Easlea, *Science and Sexual Oppression: Patriarchy's Confrontation with Woman and Nature* (London: Weidenfeld and Nicholson, 1981), 73.

25. Vandana Shiva, "Reductionist Science as Epistemological Violence," in *Science, Hegemony, and Violence*, ed. Ashis Nandy (Delhi: Oxford University Press, 1988), 232–56, 246.

26. K. M. Tewari and R. S. Mathur, "Water Consumption and Nutrient Uptake by Eucalyptus," *Indian Forester* 109 (1983): 851, quoted in Vandana Shiva, "Reductionist," 245.

27. Vandana Shiva, "Reductionist," 237.

28. Ibid., 232.

29. F. Troup, *Forbidden Pastures: Education under Apartheid* (London: International Defense and Aid Fund, 1976), 8.

30. Quoted in B. Rose and R. Tunmer, *Documents in South African Education* (Johannesburg: A. D. Donker, 1975), 205.

31. Quoted in Rose and Tunner, *Documents*, 213.

32. A. N. Mbere, "An analysis of the association between Bantu Education and Christian nationalism: A study of the role of ideology in education." Unpublished doctoral dissertation, Harvard University, 1979.

33. Simphiwe Hlatshwayo, *Education and Independence: Education in South Africa, 1658–1988* (Westport, Conn.: Greenwood Press, 2000), 74.

34. W. Rodney, *How Europe Underdeveloped Africa* (Washington, D.C.: Howard University Press, 1974), 240.

35. M. Horrell, *A Decade of Bantu Education* (Johannesburg: South African Institute of Race Relations, 1964), 167.

36. Quoted in L. Chisholm, "Redefining skills: Black education in South Africa in the 1980s," in *Apartheid and Education*, ed. P. Kallaway (Johannesburg: Ravan Press, 1984), 387–410, 389.

37. Martin Heidegger, "The Question," 27/"Die Frage," 30.

38. Quoted in Troup, *Forbidden*, 20.

39. Hlatshwayo, *Education*, 20.

40. S. Bowles and H. Gintis, "Schooling in Capitalist America: Reply to our Critics," in *Bowles and Gintis Revisited: Correspondence and Contradiction in Educational Theory*, ed. M. Cole (London: Falmer Press, 1988), 2.

41. Rose and Tumner, *Documents*, 232.

42. Kristen Shrader-Frechette, "Reading the Riddle of Nuclear Waste: Idealized Geological Models and Positivist Epistemology," in *Earth Matters: The Earth Sciences, Philosophy, and the Claims of Community* (Upper Saddle River, N.J.: Prentice-Hall, 2000), 11–24, 12.

43. Kristen Shrader-Frechette, *Burying Uncertainty* (Berkeley, Cal.: University of California Press, 1993), 4.

44. Kristen Shrader-Frechette, "Reading," 15.

45. Ibid., 16–18.

46. See especially *Unequal Protection: Environmental Justice and Communities of Color*, ed. Robert D. Bullard (San Francisco: Sierra Club, 1994); and *Environmental Justice: Issues, Policies, and Solutions*, ed. Bunyan Bryant (Washington, D.C.: Island Press, 1995).

47. See *Faces of Environmental Racism: Confronting Issues of Global Justice*, ed. Laura Westra and Peter Wenz (Lanham, Md.: Rowman and Littlefield, 1995).

48. Andrew Feenberg, *Alternative Modernity* (Berkeley: University of California Press, 1995), 144–66.

49. Ibid., 149.

50. Ibid., 152.

51. Ibid., 160.

52. Ibid., 159.

53. Andrew Feenberg, *Questioning Technology* (New York: Routledge, 1999), 127.

54. Ibid., 142.

9

The Human Condition in the Age of Technology

Gilbert Germain

Will Earth soon become humanity's phantom limb?

—Paul Virilio

I

Is globalization a problem? Is our understanding of what it means to be human challenged by the fact that where one lives on the planet is less determinant of how one lives today than ever before? I submit that the answers are both yes, but not simply because the planetary embrace of modern technologies appears to bring with it a convergence of cultural, economic, and political practices. There is more at stake here than the prospect of increasing societal homogenization, for globalization promises more than mere sameness. I want to press the additional claim that the technologies responsible for the shrinking of the globe are by definition also responsible for humanity's unprecedented release from the earthly constraints of space and time, constraints that traditionally have been central to our understanding of what it means to be human.

It is axiomatic that human existence is played out within the context of the earth. It is into this material order, an order not of our making, that we are born, and it is from this same realm that we depart upon our death.[1] Moreover, as beings who inhabit the earth, we are also of the earth to the extent we share with the world the properties of spatial and temporal extension. We could conclude, then, along with Hannah Arendt, that the earth is the "very quintessence of the human condition."[2] Yet, perversely, we humans always have been uneasy with our status as earthly beings. Max Weber has noted that, historically, we have sought to mollify this existential unease by diminishing the otherness of our worldly environs. We want our earthly "house"—which has been given to us, ready made—to be a home, to bear the imprint of our presence. According to this reading, technology is the means by which we moderns have sought to make a home of the world, as magic was the means for premoderns.[3]

Domesticating the earth through technology is problematic, however. For there is a difference between making a home of the world, and remaking the world so thoroughly in an effort to humanize it that "the world" effectively ceases to be, along with its humanizers. The latter route is the one we appear to be following, and it speaks to our radical disenchantment with the given.[4] Ultimately, it is the disavowing of humanity's earthly ties that propels modern technology in its quest to reconcile self and world by substituting a world of human making for the given world. While this project is tacitly justified on the grounds that the technological reordering of the given will make a home out of a mere house, I will argue here that we are ill-fitted for this new abode and that finding in it a true home would necessitate nothing less than a relinquishing of our humanity.

II

When Arendt first underscored the centrality of the earth to the human condition almost half a century ago, it was in response to her concern that there was a developing "loss of contact between the world of the senses and appearances and the physical world view. . . ."[5] She contended that modern science had opened a rift between the world we live in and the world we know. In the former world, Arendt explains, we are certain "that what we perceive has an existence independent of the act of perceiving." She goes on to assert it is precisely this perceptual faith that modern science undermines in its pursuit of "true reality."[6]

Our faith in the congruence of the worlds of experience and knowledge was dealt a fatal blow early in the twentieth century with Werner

Heisenberg's articulation of the uncertainty principle. It was Heisenberg who informed us that the notion of a really real, or objective, world, is an illusion. He countered that the physical world, at its most elemental level at least, is inconceivable as existing independently of the perceptual act, insofar as the act of perceiving and measuring "reality" alters the nature of the reality perceived. Arendt draws from Heisenberg's observation that to the extent that the subject is implicated in the constitution of the object, man effectively confronts himself alone.

For Arendt, the uncovering of the illusion of objectivity in the natural order finds a parallel within the technological order itself, of which humans are a part. The technological matrix in which we are embedded "makes it more unlikely every day that man will encounter anything in the world around him that is not man-made and hence is not . . . he himself in a different guise." The astronaut to Arendt was the "symbolic incarnation of Heisenberg's man,"[7] a being totally alienated from the otherness of the given world and thus wholly immersed in and continuous with his constructed environment. Of course, what troubled Arendt most is that, in formal terms, the fate of the astronaut is the fate of every inhabitant in a sufficiently advanced technological society: we are all fast becoming denizens of a world whose objectivity is illusory.

The arc of technological development into the twenty-first century has done little to assuage the concerns of Arendt and other like-minded thinkers. Doubtless, if she were alive today her trepidation toward the cocooning of humanity within a second-order reality would be more acute than ever. To further explore humanity's exodus from earthly reality, and the consequences of its remove, I will press into service two figures well known for their commentaries on the contemporary technological order: Jean Baudrillard and Paul Virilio.

III

Baudrillard and Virilio concur that the story of technology is the story of humanity's decreasing dependence on the earth, and that this growing autonomy from the "real world" amounts to a kind of self-alienation, as well. Despite differences in both approach and tone, their analyses corroborate the view that our drifting from the earth challenges a number of key assumptions regarding what it means to be human.

To help frame the investigation, it is important to note that the technologies responsible for distancing humanity from the earth fall into two general categories. On the one hand, the freeing of the bonds linking humans to the earth can be traced to the technologies directly effecting this dissolution. In addition, humanity's remove from the earth is enhanced by those technologies that further attune humans with their created environment. In the former instance, technology's impact upon its makers is best evaluated from the context of the earth, where the operative assumption holds that humans are earthly, or embodied, beings. Here, technology alienates humanity from its geophysical environs in a manner analogous to the way a spacecraft alienates an astronaut from the earth it orbits. In the latter instance, where the frame of reference is reduced to the created order itself, technology is assessed in accordance with the internal criterion of efficiency. The goal here is to reconcile the technological order with its human users. To continue with the metaphor employed above, we could say that we moderns are not content merely to leave the globe, but to do so in a vehicle so responsive to our willing that its status as an alien "thing" effectively disappears.

In more detailed terms, our escape from the pull of the earth is being realized along two fronts. First, we are transcending the limits of the earth by constructing a space—a "cyberspace" or "virtual reality"—that corresponds to the real world but is not of it, and is therefore unconstrained by the limitations imposed by earthly reality. This immanent transcendence of geophysical space is effected primarily by so-called new, or digital, technologies, and the virtual and simulatory "worlds" that issue from them. The technologies of alienation are counterpoised by technologies that aim not to transcend the given world as much as integrate further human beings with their environment. Technologies that seek to augment the world—both the given and created worlds— fall under this second category, the most obvious being so-called "smart" technologies.[8] The implicit goal of these technologies is to better integrate the world of atoms with the world of bits for the purpose of enhancing the efficiency of our control over the given and created realms.

IV

We are told repeatedly today that the "information age" is upon us. The industrial era, with its production of substantive consumables, is fast giving way to the production and exchange of information—to the much heralded "knowledge economy." This transition sees to it as well that the transportation infrastructures that anchored the era of industrial production are being supplanted by their information counterparts: Electronic networks of information exchange are overshadowing "real world" transport systems. These myriad pathways and communication nodes in turn have created a new kind of reality alongside earthly reality, an informational space that at the very least supplements geophysical space, or the real world.

It need hardly be emphasized that increasingly our communications with persons and the world are mediated via electronic technologies, and that this means of communication is absolutely crucial to our functioning as residents of a developed technological society. The question before us therefore pertains not to the efficacy of cyberspace, but to its suitability to embodied beings such as ourselves. We could ask: Is cyberspace just another "reality option," or does cyberspace diminish at some fundamental level the human experience?

Baudrillard's thoughts on the "obscene" offer a good start for an examination of this question. For Baudrillard, the pervasiveness of contemporary communications technologies has led to a qualitative transformation in the nature and role of the image in society. He contends that this transformation has reshaped the entire visual aesthetic, and employs the rubric "obscene" to denote this altered state. The mania for transparency and total illumination, which this term is meant to convey, radically transfigures the perceptual realm by presenting the perceiver with televisual imagery that is said to lie at "a special kind of distance" from the body that the body cannot bridge.[9] Baudrillard's argument here recalls Maurice Merleau-Ponty's observation that one can possess the visible only if one in turn is possessed by it.[10] With unmediated or natural vision, Merleau-Ponty argued, the space between perceiver and perceived is the distance separating two "objects" within a singular and shared realm of being—the phenomenal or real world of appearances. The distance between the seer and the seen is bridgeable with natural perception precisely because the two poles of the perceptual experience partake of and occupy the same world. Given this understanding, it is evident that the "unbridgeable distance" associated with virtual perception is less a true distance (which implies a common world) than a kind of schism separating two incommensurable realities—the real and the virtually real.

Baudrillard suggests that the charm of the real world issues directly from the gap that simultaneously separates and conjoins the perceiver and the perceived. It is this distance that lends reality its capacity to resist being "all actual under the look"[11] (to borrow again from Merleau-Ponty), a capacity that accounts ultimately for the elusive and seductive quality of real appearances. In conclusion, Baudrillard suggests that real or embodied perception is characterized by an active interplay between the perceiver and the perceived. The eye is seduced and brought into the world of appearing things precisely because what is "there" before the eye resists being fully revealed in a glance.

Real perception is incomplete in a way that virtual imagery is not. With the latter, what you see is what you get. Should an aura of mystery attend a perceived virtual "object" it must of necessity be a programmed aura—a contrived mysteriousness. It is because the virtual image is "all actual under the look" that what otherwise lies at a distance and is partially hidden, is brought up close and made transparent. It is this collapse of space and the attending intimacy of the virtual image that prompts Baudrillard to label it obscene. Like pornography, virtual imagery reveals too much, and in so doing appears realer than real, or hyperreal. And like pornography, this surfeit of visibility captures and fascinates the eye.[12]

Virilio is in basic agreement with Baudrillard on this point. For instance, in *The Vision Machine* he cites approvingly Merleau-Ponty's claim that with unmediated or natural vision everything one sees is "marked on the map of the 'I can.' "[13] Such an assertion underscores what has been said previously regarding the linkage between natural perception and real space. To argue as Merleau-Ponty has that with natural vision seeing and doing overlap is to say that the perceiver and the perceived occupy the same world. The assumption here is that as an embodied perceiver one has the power to traverse the distance separating oneself from the "other." This assumption in turn presupposes the even more basic conviction that both the seer and the seen inhabit the same ontological space.

For Virilio, so-called "distance technologies" undercut the sensibility that takes earthly reality to be the common space within which the perceiver and the perceived are embedded. How this sensibility is subverted is best illustrated by revisiting our discussion of the astronaut. Earlier we cited Arendt's comment that the astronaut can be regarded as a modern-day symbol of a being isolated from the otherness of the given world. The implication was that the astronaut's "home"—his spacecraft—was wholly a product of human artifice, and thus totally removed from the given, natural world. In *Open Sky,* Virilio recounts for his readers the experiences of several American astronauts who did not merely leave the planet, but actually stepped onto the surface of another celestial body.[14] He elicits from the astronauts' descriptions of their lunar excursions that the moon for them was not merely "another" world, but, more profoundly, an altogether different reality.

What accounts for this new reality is the quality of light found on the moon's surface. Virilio sees in the astronauts' commentary proof of the correlation between light and reality.[15] Citing Apollo 11 astronaut Buzz Aldrin's description of light on the lunar surface as "weird," Virilio adds that Aldrin attributed this weirdness to the fact that, unlike conditions on earth, solar light remains unrefracted by an atmosphere before hitting the moon's surface. Aldrin observed that the harshness of the moon's unrefracted light was especially evident when he moved his hand from the shadow to the light, a process he described as akin to "crossing the barrier to another dimension." The "super-bright spotlight" that was the sun also seems to have wreaked havoc with Aldrin and others' ability to estimate accurately distances and gradients on the lunar surface. Evidently, the ease and speed with which their bodies adapted to the new set of demands made on them by their one-sixth gravity environment was not matched by the adaptive powers of vision.

Virilio gleans from these observations that our visual capacity to assess the body's environs is not a generic place-independent capacity, but context specific. Aldrin's comments also inform us that when vision is alienated from its originary context, as is the case on the atmosphereless moon, a different perception of reality emerges, which for Virilio amounts to the emergence of a different reality altogether. It could be said that the lunar astronauts were immersed in a "digital" or "binary" reality to the extent the moon is a world without penumbras. Aldrin perceived the transition from shadow to light as unearthly in its abruptness, with no in-between separating the presence and absence of reflected solar light. Virilio concludes from this observation that objects on the lunar surface are not illuminated by the sun as much as exposed by the sun's glare. They are either real and there in the sun's light, or they are not. While objects on earth appear to endure as they sweep from darkness to light and back again, on the moon objects commute instantaneously between lit and unlit states. It is the instantaneity of the changeover that lends a binary or digital quality to light's effects, and gives objects the appearance of flickering, quantum-like, from one dimension to another. It is the radical disparity between the lit and unlit portions of the lunar surface that also accounts for the perceived compression of distances and the exaggeration of gradients.

Lunar reality for Virilio is a three-dimensional analogue of cyberspace. The sun's light in lunar reality is for all intents and purposes the same light that exposes "light-objects" in virtual reality. The implication is that we are all astronauts to the extent we live within the glow of the unrefracted light of a computer monitor. Similar in form to the perceptual experience of a lunar astronaut, what one perceives in virtual reality is a realm of collapsed space populated by "objects" that pop in and out of view, in a trice.

Whereas Baudrillard links earthly reality to distance and Virilio to light, the difference between their respective commentaries on the unearthliness of cyberspace is very slight indeed. It is slim because, ultimately, distance and light themselves are correlated. When Virilio attempts to illustrate how reality and light are interconnected in the example of the moon walkers, he is not so much talking about light per se as the quality of light. On these terms, the difference between light on earth

and its sister planet is significant. But what truly accounts for this difference is the variance of the worldly conditions on these two astronomical
bodies. Specifically, it is the presence or absence of an atmosphere that
determines the extent to which we can accurately read the topography
of our environment. Insofar as the earth constitutes the worldly context
in which humans have evolved, it is hardly surprising that our physiology
has adapted to the conditions specific to it, and that earthly reality is the
context for which we are best suited. Clearly, the perceived weirdness of
the lunar surface indicates as much.

Cyberspace is equally weird for Baudrillard and Virilio. Whereas
Baudrillard credits cyberspace's strangeness to its distanceless spaces, and
Virilio to its exposure by light, both use the conditions of the earth as
a reference point for their respective critiques. For Baudrillard cyberspace
is weird because, unlike earthly reality, it is a substanceless realm without
extension. Likewise, for Virilio cyberspace is alien because, in contrast to
earthly reality, it is a realm whose "objects" are exposed by light.

V

For Baudrillard, the unearthliness that characterizes cyberspace is fast
becoming a general condition of the entire technological order. Our
relationship to the "real world" of technology parallels the obscene character of virtuality in that there is a progressive collapsing of the space
separating humans from their technologies. The term *space* is not to be
understood here in a physicalistic sense but as the gap between desire
and its satisfaction as it relates to the human-machine interface. Thus, it
is the space between an actor and an acted upon object that accounts for
the relative unresponsiveness of the latter toward the intentions of the
former, for the object's "unfriendliness."

Ornery hardware is a problem in need of a solution. In the words
of a prominent researcher at MIT's Media Laboratory, we humans have
a right to "use technology without attending to its needs."[16] To exercise
this right—the right to reconcile humans with their created environment—technology must be made transparent: it must disappear behind
its own utility. Baudrillard concurs that the seamless interfacing of humans and the created world is the sought-for end of much technological
development. He informs us this end is being realized today in a double
process that advances both the humanizing of our technologies and the
technologizing of humanity.

For Baudrillard, the same radical disaffection with the contingency
of appearances that incites us to (virtually) overexpose the world, prompts
us as well to reorder the given world as a manageable order. Simulation

is the blanket term Baudrillard applies to the various facets of this reconstituted, user-friendly order.[17] Although he contends that all kinds of systems (i.e., political, economic, social, etc.) are simulated today, for our purposes here I will focus on simulation as it pertains to the relationship between humans and their technologies.

Baudrillard claims that in the name of "maximizing performativity" we are ruthlessly excising those spaces, or system "noise," that obviate the collapse of subject (the human user) and object (the used artifact). The operative assumption behind this drive is that only by eradicating the sources of noise within a system can the war against the intransigent "other" be successfully waged.[18] Take, for instance, the paradigmatic case of advancing automotive technology and its impact on the experience of driving.[19] Of late, the most impressive advances in car performance have originated from the implementation of digital technologies. With the aid of on-board computers, cars routinely monitor and adjust their own operating systems as well as keep their operators informed of the proceedings. These aptly named "smart" cars are paragons of functionality because they communicate. On the one hand, they are forever "speaking" to themselves by replying to perturbations detected either within the operating system itself or between the system and its immediate environment. At the same time these automobiles speak to their drivers in the form of informational readouts on their ongoing operational status, demanding replies from their human operators only on those occasions when perturbations within the system cannot be self-corrected.

Baudrillard brings to our attention that the functioning of this and all integrated technological systems demands complicity between the machine and its human operator. Each pole of the dyad "human-machine" must read and correct the behavior of the "other" to optimize the performance of a given technology. The technology itself demands that each of its constituent elements—including its operator—perform their appointed functions so as to maintain the integrity of the feedback loops that secure the system's performance. Baudrillard notes that under these circumstances technology users cease to regard the technology in question as a means of extending their power. Increasingly, a car is less a thing that is driven, an object of our command, than a technological system requiring our measured input for its functional fulfillment. Baudrillard concludes that the distinction between cause (the human operator) and effect (the technical operation) is sufficiently blurred in any evolved technology that the operator/subject properly disappears into the ecology of the system.

Whereas Baudrillard tends to view the collapsing of spaces in terms of the subsuming of the self into the logic of putatively exogenous technological systems, Virilio prefers to see this process as a grafting of technology onto the self. We are fast becoming, in the latter's words, "the hood over the engine of micromachines capable of transplanting life and transfusing its impulses with the help of computer software."[20] The infusion of "intelligence" into our technological milieu, the consequences of which primarily interest Baudrillard, is therefore complemented by Virilio's considerations on the colonization of the self by intelligent machines.

The habitation of the body by intelligent machines Virilio calls the "new eugenics."[21] The implicit aim of this new eugenics, he says, is the making of a "metabody" through the addition of "superficial prostheses" and "intraorganic nanotechnologies." Despite differing emphases, the consequence for the individual of the grafting of technology onto the bodily self is largely congruent with Baudrillard's assessment of life in an age where persons have adapted to the demands of their technological milieu. For the latter, our succumbing to the "sacrificial religion of performance"[22] has produced a new form of individualism which bears no resemblance to its bourgeois predecessor. The "postmodern" or "neo-" individual for Baudrillard is nothing more than a proverbial cog in the machine, an interactive, plugged-in being both self-identical and at one with its created environs.

The dominant existential mood of the neo-individual can best be described as dispassionate intensity. On the one hand, the postmodern self is dispassionate because passion suggests desire, or a "space" between desire and its satisfaction, the existence of which Baudrillard says is ruthlessly being erased in a world that demands transparency. The postmodern individual is compelled to dispassionately operate a technology, such as an automobile, precisely because the technology presents itself to its operator as an interactive system whose optimal functioning (i.e., the desired end) demands human surveillance and input (i. e., the means to the satisfaction of this end). On the other hand, operating a technology is an intense experience, owing to the extensive powers of control offered by most new technologies. Yet because the considerable powers of control at our disposal are not properly ours, but emanate from a system within which we participate, this intensity is tempered or restrained. The prevailing mood of Baudrillard's new man is perhaps best exemplified by the stereotypical "thumb surfer," who intently surveys the televisual landscape in the hope of stumbling across yet another momentary diversion. This and other inhabitants of hyperreality exhibit the kind of brittle intensity that can arise only when the pursuit of an end is conflated with the extension of means.

Virilio, too, believes that closing the gap between humans and their technology is helping to create a new ethos that is at once both cool and hot, dispassionate and intense. He notes that only with the technological advances of the machine age did humanity begin to free itself from the "traditional conditions of existence,"[23] whose harshness had till then impelled us to economize physical activity in an effort to maximize long-term survival. The "law of least action" has been in decline since that time, and it is almost entirely deactivated today with the advent of distance technologies and the resulting eclipse of geophysical space. Rather than conserve bodily energy, as was necessary when our very existence relied on its expenditure, the redundancy of the human body in the information age produces the obverse need to live "more acutely."[24]

Again, intensity is the order of the day. For Virilio, the human body in a body-less world—in a world whose geophysical substance has been eclipsed technologically—is an anomaly that can be normalized only by tuning the body's rhythms with those of its electronic environment. This invariably entails the acceleration of human physiological processes, since the information age is founded on the principle of instant communication. The impetus, therefore, is to treat the body as a motor or machine that "needs to be constantly revved up," either through the ingestion of chemical stimulants or the implantation of biotechnologies that amplify basic physiological capacities.

Virilio follows Baudrillard in arguing that the intensifying of lived existence is ultimately a cool form of intensity. That is to say, the drive to overstimulate the nervous energies of the body is accompanied by a countertrend to tranquilize these same energies by managing and coordinating their expression. If for Virilio the human body is quickly becoming motorized, then like a motor the body is destined to be programmed: like any other technology, the "metabody" is a cybernetized system.

VI

It has been argued here, along with Arendt, that the human condition is defined in large part by its geocentrism. It has been argued as well that we desire to be at home in this earthly abode, to reconcile ourselves with the "free gift"[25] that constitutes both the being of the earth and ourselves as earthly creatures. Importantly, reconciling ourselves with the given involves resisting the temptation to resent what is not of our own making. Reconciliation and alienation prove not incompatible terms in this context.

To argue that one can be reconciled with an environment from which one is nevertheless alienated seems contradictory only if alienation is conflated with resentment. It is axiomatic that human beings are not simply in and of the world, as are insentient objects. So although the earth may be the very quintessence of the human condition, it does not follow that humans are at one with the earth or that they need identify with it in any simplistic sense. Indeed, reconciliation with the earth means, if not openly embracing, then at least not resenting alienation as a core constituent of the human condition. When our comportment toward the otherness of the world's materiality—including that of the earth, ourselves as corporeal beings, and even our creations—which is always marked by a certain discontentment, transforms itself into contempt, the stage is set for the overcoming of alienation and, along with it, the overcoming of both humans and world.

This lesson is not lost on either Baudrillard or Virilio. The former believes we are in the midst of a war against the intransigent "other" that cannot be won because we misguidedly assume victory can be realized only by eradicating all the sources of system noise. This strategy contravenes what may be called the wisdom of superfluity,[26] a central organizational principle of both the natural and social orders that ensures a place for redundancy within the economies of these orders. Given this understanding, Baudrillard postulates that the difficulties associated with postmodernity originate "not in an excess of alienation, but in a disappearance of alienation in favor of a maximum transparency between subjects."[27]

The message is clear for Baudrillard: "Total prophylaxis is lethal."[28] By maximizing the interconnectedness of a system's elements, which of necessity entails collapsing the space separating these elements, a "malignant reversibility" is generated that destabilizes the entire system. In short, there is a point beyond which the maximizing of efficiency is self-defeating. No one, it turns out, is more vulnerable to catastrophic reversals and collapses than Arendt's "astronaut" or Baudrillard's "Boy in the Bubble."[29]

Virilio likewise regards the disappearance of alienation with trepidation. Technology, he says ironically, is well on the way to putting an end to "the scandal of the interval of space and time that used to separate man so unacceptably from his objective."[30] Virilio can make such a claim because he works from the premise that the materiality of the "space-world" (i.e., geophysical reality) acts as a break in space and time, lending to both the quality of extension. So it is, for instance, that in the real world of appearances an object (including the human body) cannot exist in two different places simultaneously. If an object does occupy two sites, common sense tells us an interval of time must have elapsed between these positionings. Virilio concludes that time is nothing other than "the form of matter in motion."[31] Moreover, because in the context of real phenomena speed pertains to objects with mass, speed in its worldly register always falls short of its upward threshold—the speed of light. Thus, the space-world proves to be speed's nemesis in that it withholds from speed the promise of its perfection. And it is this withholding that is crucial for Virilio, for the exaction of the payment of time demanded by traversing the terrestrial surface accounts in part for the depth of embodied experience.

It is because we resent earthly reality, taking it to be a drag—quite literally, as responsible for the dragging out time and space—that the refractory and hence alien space-world is being supplanted with the resistantless light-world called cyberspace. In this world where the "scandal" of space has been obliterated, durational time *(durée)* is superceded by "real-time," and the "depth of historical successivity" has yielded to the "relief of instantaneity."[32] Virilio takes as proof of the earth's demise that what matters most in our digital age is not the actuality of an event, but "being there" as it happens. We value the immediacy of real-time interaction with virtual representations of events over real-world participation in the concrete events themselves.[33]

VII

The ongoing digital and communications revolution continues to compress both geophysical space and the space separating humans from their creations. By dematerializing the real world of appearances and embedding ourselves seamlessly within a world of our own making, we are more than ever liberated from the "body" of the world and our own somatic selves. Should these developments be a matter of concern? Absolutely, and for at least two reasons. On the one hand, our liberation from the materiality of the world is purchased at the price of inhabiting a parallel world of incomparably less depth and density. While the virtually real world is habitable, in some sense of the word, it remains ultimately a charmless realm devoid of the existential heft, and thus the enduring allure, of the world it progressively occludes. In addition, because we are not beings who merely happen to have earth-inhabiting bodies but beings for whom earthly embodiment is the sine qua non of existence, the ascendancy of the virtual over the real undercuts the very reality of human existence.

The interpenetrating of human and machine only reinforces the unreal or abstract quality of much of contemporary existence. As our earthly home is radically transformed into a singularly human environment, whose artifice means we confront nothing but ourselves, we risk losing ourselves in our self-reflected image. It is the space, or tension, between humans and their recreated environment that keeps the poles of the dyad from imploding and forming a hybrid that is neither strictly human nor machine.

Specifically human aspirations and realizations heretofore have emerged in the relation to the limits imposed on us both by our given nature and the natural order itself. By gaining control of the very substance of our being and by attuning the rhythm of our physiological processes to the tempo of the new technologies, we are not merely altering the accidental qualities of some inviolable human essence, but tampering with the very core of our humanness. The protracted attempt we are witnessing today to extricate humanity from its earthly confines speaks to our radical disenchantment with the human condition. Our embrace of the new technologies is therefore not an affirmation of life as much as a kitschified version of an existence whose end is to purge an "impure" materiality from both world and self.[34] It is a life befitting the satisfied and soulless animals of posthistory.

Notes

1. To assert that the human condition is fundamentally earthly is not to make the more extravagant and reductionist claim that the full spectrum of human experience can be adequately captured in materialist terms alone. There is no contradiction contained in the assertion that what is revealed to us as earthly beings may exceed the merely earthly.

2. Hannah Arendt, *The Human Condition* (Chicago: The University of Chicago Press, 1958), 2.

3. For an overview of Max Weber's analysis of the historical relationship between humans, nature, and technology, see Gilbert G. Germain's *A Discourse on Disenchantment: Reflections on Politics and Technology* (Albany: State University of New York Press, 1993).

4. The "we" to whom I am referring is primarily, but not exclusively, the populations of the technologically developed Western world.

5. Hannah Arendt, "The Conquest of Space and the Stature of Man," in *Between Past and Future: Eight Exercises in Political Thought* (New York: Viking, 1954), 273.

6. Hannah Arendt, *The Life of the Mind*, one-volume edition (New York: Harcourt Brace Jovanovich, 1971), 46.

7. Arendt addresses the issue of Heisenberg and his relevance to contemporary technological culture in *Between Past and Future*, 275–78.

8. The "reality" that technologies augment comprises both the realm of human artifice and the natural order. Smart technologies is a term usually reserved to describe those technological means that help further integrate humans with their created environment. Of course, the given world can be augmented (as the rubric "genetic engineering" indicates vis-à-vis the human body) as well, but its inclusion here as a topic of consideration would extend beyond the confines of the present analysis.

9. Jean Baudrillard, *The Transparency of Evil: Essays on Extreme Phenomena,* trans. James Benedict (London: Verso, 1993), 55.

10. This characterization of vision is not as fanciful as it might otherwise seem if we keep in mind that for Merleau-Ponty, perceiving, like touching, is an embodied act occurring within the world's flesh. In his effort to overturn the Cartesian understanding of vision as an internal mental representation of an external object, Merleau-Ponty reminds his readers of the basic attunement between the seer and the seen that attests to their common home in the world's flesh. Like a pair of ballroom dancers whose coordinated movements are so expertly and effortlessly executed that one cannot readily distinguish the person who leads from the one who follows, the relation between the seer and the seen is said to be "magical" in its exquisite complementarity. See Maurice Merleau-Ponty's, *The Visible and the Invisible,* ed. Claude LeFort (Evanston: Northwestern University Press, 1968), 146.

11. Merleau-Ponty, *The Visible and the Invisible,* 191.

12. Jean Baudrillard, "The Ecstasy of Communication," in *Postmodern Culture,* ed. Hal Foster (London: Pluto Press, 1985), 132.

13. Maurice Merleau-Ponty, *Phenomenology of Perception,* trans. Colin Smith (London: Routledge and Kegan Paul, 1962), 137.

14. Paul Virilio's *Open Sky,* trans. Julie Rose (London: Verso, 1997), 118–45.

15. Ibid., 138.

16. Neil Gershenfeld, *When Things Start to Think* (New York: Henry Holt, 1999), 102.

17. For Baudrillard, radical disaffection with the contingent nature of the real or unreconstructed world provokes an equally radical desire to recreate the world as a manageable order. The control that simulation promises is achieved through the employment of feedback mechanisms of various sorts. Simulation within the economic realm, for instance, is evident in the now routine practice whereby the production of an actual commodity postdates its (virtual) testing in the marketplace. Commodities are not only produced but literally conceived in the crucible called "consumer feedback." This prompts Baudrillard to comment that only in the age of simulation are commodities "conceived according to their very reproducibility" (Jean Baudrillard, *Seduction,* trans. B. Singer [Montreal: New World Perspectives, 1990]), 56. See also Jean Baudrillard, *Simulations,* trans. Paul Foss, Paul Patton, and Philip Beitchman (New York: Semiotext(e), 1983), 171.

18. Jean Baudrillard, *The Illusion of the End,* trans. Chris Turner (Stanford: Stanford University Press, 1992), 81.

19. The further integration of humans with their technologies takes on a multiplicity of forms today. Apart from some of the more obvious examples— such as smart houses or fly-by-wire aircraft—the most interesting advances in this integration process are happening in the computer technology field. For instance, The Tangible Media Group, a subdivision of MIT's Media Laboratory, has as its mandate improving "human-computer interaction" (HCI) by foregrounding the

haptic or tactile dimension of human communication. In the words of the Group, its research goal is to "rejoin the richness of the physical world in HCI." According to this reading, computers need more "body" if the goal of relating to our digital environment in a more multimodal and, hence, a more fully human way is to be realized. See Hiroshi Ishii and Brygg Ulmer, "Tangible Bits: Towards Seamless Interfaces between People, Bits and Atoms," published in the proceedings of CHI '97, March 22–27, 1997. This paper also can be retrieved at the following Internet address: http:www.media.mit.edu/groups/tangible/papers Tangible_Bits_CHI97/Tangible_Bits_CHI97.

20. Paul Virilio, *The Art of the Motor*, trans. Julie Rose (Minneapolis, MN: University of Minnesota Press, 1995), 129.

21. See Virilio, *The Art of the Motor*, 116–30.

22. Baudrillard, *The Illusion of the End*, 106.

23. In Baudrillard's words, the postmodern or neo-individual is "the purest product of 'other-directedness': an interactive, communicational particle, plugged into the network . . . ." See Baudrillard's *The Illusion of the End*, 106.

24. Virilio, *The Art of the Motor*, 123.

25. Arendt, *The Human Condition*, 2.

26. The expression "the wisdom of superfluity" is mine, not Baudrillard's.

27. Baudrillard, *The Illusion of the End*, 80–81.

28. Baudrillard, *The Transparency of Evil*, 64.

29. "The Boy in the Bubble" is Baudrillard's answer to Arendt's astronaut, a being condemned to live a life in an entirely artificial (and in this case, germ free) environment for want of a well-functioning immune system. The Boy in the Bubble is "a prefiguration of the future," according to Baudrillard, because of the reigning confusion as to what constitutes health. We erroneously assume that by cleansing ourselves and our environment of germs—the purported noise within the human biological system—we maximize our performativity, when in fact we are exposing ourselves to the "malevolent reversals" that seek out hyperintegrated systems. See Baudrilllard's *The Transparency of Evil*, 60–63.

30. Virilio, *Open Sky*, 119.

31. Ibid., 120.

32. Paul Virilio, *The Information Bomb*, trans. Chris Turner (London: Verso, 2000), 129.

33. Virilio does not look upon the collapse of both real distance and historical time with equanimity. Citing the human body as constituting the basis of his work, he makes it clear that our capacity to see anything in any place in real time transgresses the limits placed upon us as incarnate beings. In his words: "The technologies of virtual reality are attempting to make us see from beneath, from inside, from behind . . . as if we were God." See the 1994 *CTheory* transcript, "Cyberwar, God and Television: Interview with Paul Virilio," which is available at the Internet address: http:/www.ctheory.com/acyberwar_god.ht.

34. My understanding of kitsch is informed by Milan Kundera's discussion of the same in *The Unbearable Lightness of Being*, trans. Michael Henry Heim (New York: Harper and Row, 1984). See 247–48.

10

Technology and the Ground of Humanist Ethics

Ian Angus

From Bacon's *New Atlantis* to Huxley's *Brave New World*, technology has appeared alternately as savior or destroyer of the attempts to found a just and free social order. Occasionally the Faustian bargain has come to self-consciousness and it has been recognized that these optimistic and pessimistic variants rest on an identical wager—identification of sociopolitical ideals as consequences of technological development. One need only think of the 1950s, when American military and industrial technology was widely taken to be an index of moral leadership in creating an egalitarian society. In more recent times, international competition in the technological Olympics, especially from Japan and Germany, has led to some pessimism that American technology can deliver the goods in the sociopolitical arena. From Galbraith's optimism to the pessimistic followers of Jacques Ellul, the fortunes of society are taken to be determined by technology. A new version suggests that the Internet will lead us ineluctably into a decentralized, participatory society. Manifold variants of technological determinism arise automatically from whatever the prevailing conjuncture of technological development.

However, the violent social and ecological "side-effects" of technological "advance" have also motivated a more thoughtful reconsideration of determinism. Under the slogans that technology is "merely a means" and "occurs within a social context," a humanist revival has attempted to reestablish the independence of ethico-political ideals. The goals of a free and just society are to be rescued from the great tradition of Western thought, in opposition to contemporary cynicism, to provide the measuring rod whereby technology can be evaluated in its contribution to the realization of ethical ideals. Humanism consists in the defense of ethical discourse as separate from and uncontaminated by the technological extension of power.

The contemporary explosion of technology has encouraged a great deal of concern, but very little of it raises fundamental questions. The threats to employment, privacy, ethnic, regional, or national autonomy, artistic integrity, and so forth are legitimate concerns in the face of the massive social organizations and concentrations of power and knowledge that result from present technological development. But what is it that is being threatened? The right to make a living, privacy, autonomy, democracy, the possibility of individual participation in social and political decision making—all of these are important. But fundamentally, what is at issue is the humanist concept of the subject which is defended in our philosophical tradition and with which most of us operate without reflecting on its origins or contemporary viability. It might be suggested that what is required are descriptive empirical studies of the impact of technologies in various social settings. Informative as such studies are, limiting one's interest to them implies that technology poses no threat to our received ethical tradition. The utilization of descriptive accounts in judgment or action affecting technological impact requires an ethico-political evaluative component that is not simply descriptive. To suppose that descriptive accounts are sufficient is either to accept technological determinism or to presuppose that the ethical and political categories from the received philosophical tradition are adequate to evaluate technology—that the evaluative principles of equality, prosperity, privacy, autonomy, justice, democracy, etc. are not undermined by contemporary technology. Pointing to the social context of technology is generally taken to settle the ethical question. Against this, it is suggested here that humanist ethics incorporates negative presuppositions about technology that implicate its ethical categories in present technological development. Criticism of humanism is a difficult and dangerous enterprise; one must be careful not to give aid to technocratic tendencies of a far more vicious and manipulative bent. However, if it is indeed the case, as I argue, that humanist ethical assumptions are intertwined with technological assumptions, humanism cannot provide an adequate evaluation and critique of contemporary technology. Moreover, it cannot provide us with any guidelines for an alternative development. On these grounds, the task must be attempted.

The rationality that has culminated in contemporary technology originated in the Renaissance connection of human good and human power. Descartes and Bacon are both historically significant and symbolic figures in this respect. In the *Discourse on Method,* Descartes expressed his expectation that the new utilitarian knowledge would eliminate labor, conserve health, and perhaps even prevent the impairments of old age.[1] Bacon, in the *Novum Organum,* articulated the connection between scientific methodology and the domination of nature: "Human knowledge and human power meet in one; for where the cause is not known the effect cannot be produced. Nature to be commanded must be obeyed; and that which ill contemplation is as the cause is in operation as the rule."[2] One misses the connection of power and good if the problem is posed as the opposition of technology and ethics. The conceptual basis of the new science can be indicated with reference to its inclusion of both mathematics and technique. A new interest in instruments used by craftsmen and in new devices for altering perception, such as the telescope, was characteristic of the founders of modern science. Also, this technical interest coincided with the introduction of mathematics as the fundamental language of the sciences. Through this reshaping of scientific method, knowledge of nature was expected to produce potent effects, which could be utilized to realize human ends. Technical applications predictably followed the course of the new science because they were inherent in its conception of knowledge from the outset. Moreover, the mathematical substruction of nature removed any conception of cosmos or natural value-laden order from scientific theory. Consequently, technical advances could be made to serve human ends which were decided upon apart from any overarching conception of good. Liberated from cosmological restraint, human power became intertwined with humanly posited ethical goals. Due to this connection of power and good, technology must be considered to be an expression of humanism.

At the beginning of Western philosophy, Plato posed the question of how to measure or evaluate an entire culture. In *Statesman* (284d–286c) he introduced the concept of qualitative measure as pertaining to the overall human good. Quantitative measure established the excellence of a limited art to the end which it serves. For example, we measure the excellence of the art of medicine by the extent to which it promotes health. Moreover, having a skill does not mean one has the ethical insight that determines its proper use (*Republic,* 333e–334b). Qualitative measure, on the other hand, is concerned with the ends to which limited arts are directed: Are these ends conducive to the human good? Indeed, is there a conception of good that makes a claim on us as humans rather than as doctors, taxi drivers, librarians, etc.? In the *Republic* (335a–e),

Plato defined justice as the good for humans per se and concluded that it is not just to harm anyone since that would diminish the humanity of the harmed one. In order to take the measure of a culture, it is necessary to confront this question of a trans-contextual human good. Clearly, there have been many and varied concepts of human good and it is difficult to pronounce upon the intrinsically human without denying and violating other human potentialities. Yet without the possibility of measuring a whole culture, the attempt to evaluate contemporary technology is doomed from the start by cultural relativism. The contemporary revival of humanist ethics takes its sustenance from a defense of qualitative measure. It will have to be maintained, though rethought, in a criticism of humanism.

The evaluation of technological culture encounters an additional difficulty. Modern technology consists of two separate components: first, the connection of reason and technique, of mathematics and experiment, which originated in Renaissance philosophy; second, the universalization of this model such that formal reason and technical action become the only rationally justifiable types. In other words, there is a connection between formalism and technique, elements that exist in diverse cultures (though they are not necessarily linked) and are subservient to the overall mythical order justified by poetry, cosmology, religion, etc. But also, this formal-technical model has been universalized in the contemporary situation to become the defining characteristic of technological culture. It is from this unprecedented complex that the universality and rationality (i.e., non-traditionality) of modern technology derives. While machines are certainly exemplars of technological development, modern technology refers to any means to a defined end. Ends can be defined in any sphere of life once mythical world order is rejected. Since technique in this wide sense is legitimated by scientific reason, it tends to become universalized. Consequently, modern technology initiates a new situation to which the cultural measure developed in humanist ethics is not adequate.

Quantitative measure involves two objects and a standard unit by which they are brought into relationship. The standard unit is applied to each object and thereby they are rendered comparable as longer, shorter, etc. However, even in this simple case there is another element which can be easily overlooked. Not only must the correspondence between the standard unit and the object take place, it must be seen to take place. Quantitative measure presupposes a viewing subject. In the case of qualitative measure of a culture, the self that judges the overall human good becomes the measure as well as the measurer. (This is the basis for the permanent danger of misunderstanding the subject on the model of the things seen, which requires the constantly renewed task of philosophical criticism.) The possibility of a subject judging the whole culture, and participating in determining its direction, is the foundation of the humanist defense of subjectivity, which, consequently, involves both epistemological and ethical dimensions. It is on this distinction between a quantitative, limited measure and a qualitative, overall one that the humanist position stands—and, I will argue, falls.

The humanist ethical defense of subjectivity is based on two historically important distinctions in ethical and political thought—techne/praxis and hypothetical/categorical imperatives—that derive from Aristotle and Kant respectively. Consideration of these two distinctions will show how technological assumptions enter negatively into ethico-political categories. That is to say, by defining technical action initially as a-ethical, as devoid of ethical implications, humanist categories assume the ethical neutrality of technology. Consequently, humanist concepts cannot adequately be used to understand the contemporary threat posed by technology—that it has come to undermine the basis of ethical categories. In other words, the humanist subject is defined from the start as uncontaminated by technology and therefore cannot adequately criticize the ethical implications of contemporary technology. Humanist criticism is limited to the "ends" that technological "means" serve; it cannot penetrate the intertwining of means and ends—the connection of "peaceful" nuclear power and the atom bomb, scientific medicine and biological warfare, consumer goods and ecological crisis. In our time the task is to think together these kindred progeny of Prometheus and Pandora.

Aristotle refined Plato's concept of qualitative measure through the distinction between techne and praxis, or art, technique, and practical action. Some activities aim at particular ends; there are many of these. Often they are subordinated as means to ends of a more comprehensive scope. However, unless this end/means/end system is to be infinite, and thereby lack a basis for the determination of the contribution of particular ends to the whole of human life, there must be another type of activity which is an end in itself. Aristotle wrote that the most honored capacities, such as strategy, household management, and oratory, are contained in politics. Since this science uses the rest of the sciences, and since, moreover, it legislates what people are to do and what they are not to do, its end seems to embrace the ends of the other sciences. Thus, it follows that the end of politics is the good for man.[3]

The distinction between art and practical activities turns on the nature of the end in each case. Arts produce limited ends in which the end is distinct from the process of its production; thus, they do not themselves determine the ultimate utilization of their particular ends. On the other hand, practical action is its own end and determines the utilization of particular ends.

In Aristotle's view, limited technical ends can only be evaluated on an ethical basis in which interhuman action is its own end. In other words, intersubjectivity, which is the foundation for ethics, is in no way accomplished or affected by technical action. Isolated limited ends are presupposed to have no effect in determining intersubjective action. Happiness has a determinate content for Aristotle. It consists of using limited ends (such as wealth, horsemanship, or skill in crafts) for the practice of virtue in noble and good deeds which are complete in themselves and consequently even more durable than science.[4] While one may not be able to be happy if one does not have the means (e.g., wealth) to do good deeds, there is no suspicion here that the deeds one can do are constructed in type and in style, in form and in content, by the available means. But it is of little use to donate hard currency to poor who are without access to an exchange market, or goats to a modern charity that must transport the gift across thousands of miles. More generally, the goodness of a deed for Aristotle consists in its character as action, on the fact that it is seen and remembered by others. Technical means are excluded from ethical action because goodness resides solely in the intersubjective visibility of the act.

Kant distinguishes between technical, pragmatic, and moral imperatives of action. Technical rules of skill are innumerable insofar as action is necessary to bring about a practical purpose. In technical imperatives the rationality or goodness of the end is not questionable, but only the means to attain the end. Imperatives of skill are hypothetical since they prescribe what rules must be followed if a certain end is adopted but do not question whether the end is or should be adopted. The end remains unquestioned in hypothetical imperatives. Counsels of prudence are pragmatic; they are means to the end of happiness. This end can be presumed actually to exist in all individuals. However, actions of this type are also hypothetical since they are still means and the actions are not good in themselves but only relative to the end of happiness. Kant further subdivides prudence into worldly wisdom and private wisdom. The first consists of skill in influencing others to one's own ends, the second in combining one's own ends to lasting advantage. Kant notes that without private wisdom, worldly wisdom is more accurately called cleverness and astuteness but not prudence. This indicates that prudence consists in combining ends in pursuit of happiness and that an unordered collection of ends, however clearly pursued, does not tend to produce happiness. In contrast to Aristotle, Kant argues that the attempt to harmonize ends within the self to achieve happiness suffers from our ignorance:

If it were only as easy to find a determinate concept of happiness, the imperatives of prudence would agree entirely with those of skill and would be equally analytic. For here as there it could alike be said "Who wills the end, wills also (necessarily, if he accords with reason), the sole means which are ill his power."[5]

Consequently, our ideas of happiness often conflict with themselves and pose the problem of achieving a unified and harmonious concept. Nevertheless, if it were possible to describe the content of happiness, it would be possible to describe all means toward it as technical. Since omniscience would be required to decide with certainty about happiness, our decision can only rely on empirical counsels and not on determinate principles. The unification of diverse ends that happiness demands is left to experience and deemed incapable of rational judgment. The distinction between technical and pragmatic imperatives appropriates Aristotle's dichotomy of techne and praxis, with one important qualification. Like Aristotle's account of techne, Kant's discussion of hypothetical imperatives attempts to show that an adequate moral law cannot be a means to an end but only an end in itself. However, Kant rejects Aristotle's notion of a determinate concept of human happiness. The watershed between Aristotle and Kant is the Renaissance mathematical science of nature which rejects cosmological intuition in favor of the categorical determination of nature. In other words, Kant's recognition of the active nature of knowledge requires a rejection of deriving a moral law from cosmological intuition. Happiness becomes indeterminate when knowledge becomes entangled with power.

The third imperative of action is a moral imperative and is termed "categorical" since it does not depend on a prior hypothetical "if" that is outside of rational discourse. Consequently, the categorical imperative is universal since it applies in every case whatever one chooses to adopt as ends. Moreover, since there is no determining condition—such as the prior choice of an end for a hypothetical imperative—the conformity of the maxim to the law or conformity of the principle of action to the condition of universality is all that is conveyed in the categorical imperative. Kant states: "Act only on that maxim through which you can at the same time will that it should become a universal law." If one grants that a rational being exists as an end and not merely as a means, then an alternative and equivalent formulation can be given: "Act in such a way that you always treat humanity, whether in your own person or in the person of any other, never simply as a means, but always at the same time as an end."[6]

Kant's categorical imperative is an attempt to formulate a universally binding moral law for the sphere of interhuman action independent of any determinate concept of happiness. In so doing he opposes his ethic of autonomy to Aristotle's heteronomous or worldly ethic of happiness. Kant attempts to forestall the universalization of technical ends by finding in the sphere of human interaction a moral law that elevates human selfhood to an end in itself. The moral will is determined only by the condition of universality and not by any worldly content. This notion of enlightenment proposes an anthropocentric solution whereby the world of use objects and technical means are ultimately for the fulfillment of inter-human life, which is ruled by the moral law.

The choice of Aristotle and Kant to illustrate the basis of a humanist ethic illustrates a remarkable continuity throughout ancient and modern, Greek and Christian, heteronomous and autonomous traditions. Despite the diversity of various humanistic ethical systems, these two examples illustrate how humanistic ethics begins from a distinction between technical, limited, hypothetical ends, and ethical action, which is an end in itself and resides solely in the interhuman, social sphere. This distinction is equivalent to the claim that technical action is ethically neutral or, in another formulation, that technology cannot come to influence or determine the content of ethical action. Against this claim the following considerations suggest that technical and ethical action are necessarily interwoven.

If we define a technical problem, we must first of all isolate a single end from the unformulated practical context in which we prereflectively live. In order to focus on the necessity to fix a leak in my roof, to repair a broken lock or to correct a misunderstanding with a friend, the end that is desired must stand forth in my ongoing practical everyday life. This often happens when something ceases to work well—when water drips into my house, when I cannot get in the door, or when my normally affable friend becomes irritable. This obvious observation serves to illustrate a rather less obvious point: technique is not a separate type of action but rather an abstracting, focusing activity within the whole world of practical action. When one considers this abstracting technical activity it also becomes apparent that the focusing of my practical attention on a single end does not sever the connection that this end has to the practical context even though this connection is not explicitly present in the isolated end. Thus, for example, the problem of a broken lock refers me to a social situation in which we lock our doors when out of the house, a leaky roof to a rainy climate, and an irritable friend to our social institution of friendship. In other words, the socially, historically, and naturally defined situation enters into the abstraction of a technical end and serves to define when the technical task is completed. On this basis it is apparent that the contemporary explosion of technology does not consist merely in the expansion of the sphere of things regarded as "simply means" but rather in the practical context from which techniques are defined. More accurately, it is a specific configuration of the practical context that lies behind and motivates the explosion of technology. Consequently, the specificity of contemporary technological society is not to be sought in the mere utilization of techniques (which is common to every society) but in the socially and historically determinate process of abstraction in which technical ends are defined.

Another consideration pertains to the manner in which technology alters our experience of the world. If I replace my pen and write with a typewriter, or a computer terminal, the greater speed of the instrument and altered possibilities for revision encourage a different writing style— there is less of the room for thought that an aching hand introduces. Similarly, a telescope magnifies the surface of the moon only at the price of a loss of awareness of one's distance from it and of its place in the night sky. The telephone transmits speech over great distances by losing the connection to the gestural basis of communication in the human body. To conclude, technology does not merely add to our experience but alters it and directs our attention to certain aspects of the world— one is directed to either the forest, the trees, the leaves, cells, or atoms. Technological alteration of experience inclines us toward the phenomena which they exaggerate and tends to downplay what the technology is disinclined to present.[7]

These considerations indicate that the humanist account of technical action, which sets it aside as without ethical import in favor of strictly interhuman action, is no longer tenable as the basis for an ethical critique of technology. Instead, it is necessary to understand technology as an abstracting of limited ends within a practical context and as a directed alteration of experience within that context. Contemporary concern with the impact and evaluation of technology requires a new ethical foundation that centers on the interweaving of technique and ethical action. The humanist concept of the self, which is present in quantitative or technical measure and which itself becomes the standard in qualitative measure of a culture, is inadequate for the contemporary phenomenon of technology. The humanist self sees a quantitative measure to hold: it is not formed or altered by the measuring process. Consequently, ethics concerns the seeing subjects alone and is entirely a-technical. The present account of technology emphasizes the defining of the practical context performed by technology and the directed alteration of experience that it entails. On this account, it is apparent that the subject is bound up with the defining and altering of experience performed by technology.

Technical "means" come to define the content of the subject, the definer of "ends," which is the presupposition of humanist ethics.

This assertion—that the content of ethics is defined and altered by technology—must be clarified. On the one hand, it means more than the traditional recognition (quite compatible with humanism) that one's capacity to actually achieve ethical goals is limited by the technical means at one's disposal; for example, that the just distribution of wealth requires a sufficient amount of wealth and an adequate distribution system. In this case, technology is taken to be important for the realization of an ethical goal, but not for the definition of the goal itself. On the other hand, the assertion does not mean a total abandonment of ethical evaluation to technical possibilities altogether—although this is indeed the danger that technology poses in our time. What are at issue here are the presuppositions concerning subjectivity that underlie ethical action. It has been pointed out that both Aristotle and Kant begin the discussion of ethics by separating an interhuman sphere of action from the whole context of human life. On the basis of the contemporary understanding of technology outlined above, this separation can no longer be regarded as tenable. Aristotle regarded art, techne, as a limited end and sought to exclude technical acts from the institutionalized sociopolitical sphere. Kant was less concerned with the institutional location of these activities since the Renaissance watershed disclosed them to be subject to human autonomy. However, he determined the principle of ethical action in its abstraction from all hypothetical ends—ends that one may or may not choose. In both these accounts, technique is relegated to the realm of contingent choice without ethical content; ethics consists in universal and necessary principles of interhuman action.

The foregoing sketch of technology can be concretized with respect to this point. If technique consists in abstracting an end from the practical context of human life, then this context (social, historical, and natural) is non-explicitly retained in the technical end. Consequently, the contingent, limited choices that are made in pursuing actions based on technical abstractions select certain aspects of the practical context for attention and emphasis. The technical component of action determines both the selection of figures from the practical world and what recedes into an unformulated background. The social world cannot be known separately from its articulation in technical ends. Thus, the possibilities for fulfilling human power released in genetics, nuclear energy, or even as simple a case as the telephone, mute other possibilities—our choice between given technical ends is made by a subject whose attention has been riveted by the prior selection of ends. The second aspect of technology—its alteration of the temporal, perceptual, and social constitution of the practical world—also affects the ethical subject. The mode of interaction of human subjects through telephone, writing, assembly lines, knowledge factories, etc. alters the capacities of the subjects that are developed and that can be recognized by others. In short, a contemporary understanding of technology reveals the abstraction of an interhuman world from the practical world selected and altered by technology to be based on an insufficient analysis of technology. In both the selective articulation of experience and its alteration through innovations, technology is an integral aspect of the human subject. It is not that technology requires us to choose this over that value; rather, it cuts the ground out from under the subject presupposed by humanist ethics.

Modern technology is both universal and rational due to the connection of technical action to formal scientific reason. It is unlimited in scope and liberated from any overarching conception of good. The human subject can no longer be extricated by defining it as universal and necessary in contrast to the contingent and limited status of technique. Thus, the attempt to evaluate technology encounters a radical dilemma: If the self is altered by technology, how can technology be measured? How can technology be evaluated if there is no standpoint outside technology that can establish the ethical value of the self? The Renaissance connection of power and good has doubled back on the self that it was intended to serve. Technical power was to serve human ends; when technology comes to determine the content of human ends, human good is reduced to power. Technology cannot be effectively criticized by humanism because it is an expression of humanism. The supposed ethical neutrality of technology has come to sever the humanist self at its base.

This dilemma can perhaps be addressed by pointing to an ambiguity in the account thus far: technology is an expression of humanism, but technology has come to sever itself from the presuppositions of humanistic ethics. Similarly, the origin of modern technology has been located in the Renaissance connection of power and good, but humanism is taken to consist in the claim that technology is ethically neutral. In this dilemma we do not encounter merely verbal contortions. Rather, this dilemma stems from the extreme difficulty of attempting to think the implications of our present age characterized by technology. On the one hand, in automated production, scientific experimentation, computerization and, perhaps most clearly of all, nuclear energy a threat is widely, and correctly, perceived to the humanist principles that are the proudest products of our civilization. On the other hand, as I have argued above, it is less clearly appreciated that the current development of technology requires a thorough rethinking of the presuppositions of humanist ethics.

The dilemma derives from the universal and rational character of modern technology due to its basis in formal scientific reason. Consider the various levels of infinities described in contemporary mathematics. In the face of such unprecedented universality it is hopeless to try to "contain" technology and subject it to a "wider" humanistic concept of reason. In fact, it is the universalistic claims of humanism that were the basis for scientific universalism. The dilemma encountered in evaluating technology can only be approached from another angle: it is the concrete, the particular, that needs to be understood. This new angle has been indicated by the description of technology as a directed alteration of experience within a practical context. The subject is implicated in the contingent abstraction of technical ends. Universalization of technology, by which our society is distinguished from premodern and extramodern alternatives, consists in the suppression of these characteristics: scientific legitimation of technology attempts to standardize context and to ignore the direction of alteration. In other words, science abstracts from the "here and now"; it is precisely the lived dimension of this abstraction that is the specific configuration behind the contemporary universalization of technology.

"Living scientific abstraction" means treating one's "here-and-now" as simply arbitrarily designated within a formal-mathematical framework. The intellectual abstraction from one's particular characteristics has become a principle of social existence—we live as strangers to ourselves. In this sense, science and technology have become ideology: unrestricted extension of techniques legitimated by scientific reason conceals the embodied, localized practitioner, which it nevertheless presupposes. Universalization of technology conceals the concrete and particular, which it nevertheless transforms. In this connection Husserl's monumental study of formal-mathematical abstraction, *Formal and Transcendental Logic,* is of immense significance. It shows that even in the most rigorous formal ideation there are "presuppositions of sense" that determine the applicability of abstractions to concrete situations. From this angle one can begin to undo the suppression of context and direction that technology requires. This is a rediscovery of the embodied and sentient here-now which grounds the self, not apart from, but implicated in the alterations of technology.

In the attempt to evaluate technology we have been driven to recognize that our principles of evaluation are altered by technological development. The rethinking required by technology cannot remain concerned with specific impacts but must question the technological culture itself. Technological culture consists in two related phenomena originating in the Renaissance: the development of mathematical-technical science and the universalization of this method as the only justifiable form of knowledge. Consequently, an important clue to rethinking technological culture can be found in Edmund Husserl's critique of Descartes. Through systematic doubt Descartes sought to anchor mathematical natural science in an indubitable foundation in the psyche. However, in so doing, he assumed the psyche to be of the same order as the mathematically understood physical such that the validity of natural science could be derived from psychic indubitability. As Husserl put it, the psyche as understood by Descartes is misrepresented to be "a little tag-end of the world" from which one attempts to "infer the rest of the world by rightly conducted arguments according to principles innate in the ego."[8] In rethinking technology we are forced to reconsider the ground of the humanist concept of subjectivity and its fateful interweaving with the technological assumptions of mathematical natural science. Unloosing ourselves from the mathematical-physical misunderstanding of the psyche requires a return to the stream of conscious experience without assuming a privileged place for mathematical extension from which to explain other phenomena. The way in which sound, touch, or smell present the world to us must be recognized as equally fundamental as extension and sight. They are all grounded in an embodied consciousness localized in the here-and-now.

The return to embodied and sentient consciousness rooted in a practical context and formed in the directed alterations of technology seeks to undo the standardization of contexts that lies behind the universalization of technology. This return is effected in two extremes: unveiling the presuppositions of formal-universal abstraction and rigorous analysis of the reorganizing effects of various techniques. From this perspective, formal-technical science appears as one possibility (one that suppresses its own presuppositions), rather than the only possible mode of knowledge and activity. However, in this return, the specter of relativism cannot be avoided. With the emphasis on practical context and alterations of perception, with the rejection of the humanist self as epistemological and ethical anchor, how is it possible to avoid a thoroughgoing situational boundedness of knowledge and action? And if it is not possible, the attempt to measure technological culture falters. I think that this is the source of the pervasive relativism in which our society presently founders. In recoil from the stiff universality of technology, many embrace a slack, context-bound thinking. This is another way of expressing the dilemma: formal-technical science has become dislodged from its humanist origins, and these cannot be recovered in a pervasive relativism. In either case, the humanist self is on shaky ground indeed. In the short term, one can seek respite by regarding the self—its origin, growth, and protection—as "self-evident." But it is actually a hard-won cultural product, and the long-term interests of civilization are not served by this expedient. The specter of relativism is not easily exorcised. However, it should be noted that the "merely particular" is the other side of the coin of "formal universality." Formal-mathematical abstraction designates objects as numbers or, more properly, uninterpreted signs. Consequently, all content is regarded as irretrievably particular, without significance beyond itself. In other words, the predominant relativism is simply a recoil from technology, its shady side. What is needed is a radical rethinking of the particular-universal relations that have been bequeathed to us by the Renaissance under the influence of formal-technical science. Obviously, this problem can only be mentioned in this essay. Nevertheless, there is a clue to this task in the alteration of the practical context by technology. The unprecedented possibility, raised by nuclear technology, of the destruction of terrestrial life by human action, was mentioned above. There are many other examples. The techno-bureaucratic possibility, raised by Nazi total mobilization, of the elimination of Jews from Europe, made a possibility from what was previously a deranged fantasy. All possibilities are not this terrifying; however, these examples illustrate the extreme possibilities engendered by modern technology. The development of contraceptive technology has engendered the possibility of separating sex from procreation. In so doing, it has altered but not entirely overturned the prevailing relations between the sexes. The disproportion between this new technological possibility and the practical context of sexual and personal expectations accounts for the ambiguity and stress of this possibility, particularly from the standpoint of women. Technology consists in the transformation of fantasies, speculations, into possibilities—possibilities that do not leave the foundations of selfhood untouched. Herein lies the "realism" of much utopian literature.

In this notion of possibility we can begin to glimpse an approach to culture that regards the various alternatives as revealing aspects of human selfhood. From the standpoint of formal universality, for a characteristic to be inherently human, it must be empirically present in the various human cultures. Lack of this characteristic in all cultures relegates it to a merely particular, relative status. This is the source of the continual vacillation between universalism and relativism in the human sciences. If, however, we regard culture through the prism of possibility, we are concerned with the way particular cultural practices reveal possibilities of selfhood and human association.

Moreover, the humanist stress on the social context of technology ignores a fundamental phenomenon that can be clarified through the notion of possibility. Although technical ends are defined from a practical context, ends are not confined within the context in which they arise. Ends are re-contextualized; means are put to alternative use, such as when a screwdriver is used to open a paint can or pliers to hammer in a nail. The practical context is defined through the possibilities that technique focuses. This focusing illuminates the whole of the practical context, not merely the occasion in which it initially occurs. Indeed, transfers of context—such as the application of a linguistic distinction to the study of mythology by Lévi-Strauss, the architectural distinction of base-superstructure to socioeconomic phenomena, or a coat hanger to opening a locked car—appear to be characteristic of creative, metaphorical thought. The technological suppression of context has rebounded in a humanist defense of it. Nevertheless, simple allusion to the social context does not serve to understand technology any more than it solves the ethical question. Conceptualizing technology as possibility draws attention to the phenomenon of re-contextualization in which the practical context is altered by an unexpected utilization of technical means.

From these intimations we can begin to discern the contours of the fundamental dilemma in the twilight of humanism: how to recognize the universalization of contemporary technology without falling prey to it. Various formulations of this dilemma have been given. Most fundamental is the connection of power and good, which has included the claim that technology is ethically neutral, that is, is merely an accumulation of "means" to "ends" acceded to upon other grounds. The separation of technology from social interaction that characterizes humanism is inadequate to comprehend the threat posed by technology in our time. It is based on long-standing assumptions about technology, which need to be replaced with an unprejudiced phenomenology of technology. The present essay does not attempt to solve the many problems raised in this connection, but to describe the radical thought required to formulate them correctly. Technology cannot be opposed to social interaction; it is, rather, the means of accomplishing social interaction—the concrete focusing that sets up modes of intersubjective relationships. In the present, technology raises the extreme possibility of the disappearance of the self, the knowing and acting subject which is the ground for the idea of responsibility. With the understanding that the utmost possibility of technology is the severing of the subjective basis of knowledge and action, the relationship of power and good can be reformulated: the revealing of human possibilities in the variety of cultural practices can ground a new universal reflection on embodied and sentient consciousness. Vibrant within the encultured body, this reflection begins and ends with the idea of self-responsibility.[9]

Notes

1. René Descartes, *Discourse on Method,* trans. Laurence J. Lafleur (Indianapolis: Library of Liberal Arts, 1956), 40.

2. Francis Bacon, *Novum Organum*, Aphorism iii, included in Edwin A. Burtt, ed., *The English Philosophers from Bacon to Mill* (New York: Random House, 1939), 8.

3. Aristotle, *Nichomachean Ethics,* trans. Martin Oswald (Indianapolis: Library of Liberal Arts, 1962), 4 (1094b).

4. *Nichomachean Ethics,* 24–25, 257–58, (1100b), (1176b).

5. Immanuel Kant, *Groundwork of the Metaphysics of Morals,* trans. H. J. Paton (New York: Harper and Row, 1964), 85. In general, this discussion relies on pp. 82–88.

6. *Groundwork of the Metaphysic of Morals,* 88 and 96.

7. See Don Ihde, *Technics and Praxis* (Dordrecht: D. Reidel, 1979).

8. Edmund Husserl, *Cartesian Meditations,* trans. Dorion Cairns (The Hague: Martinus Nijhoff, 1969), 24.

9. This essay is a revised version of Ian Angus, "Reflections on Technology and Humanism," *Queen's Quarterly* 90:4 (Winter 1983).

11

Recomposing the Soul

Nietzsche's Soulcraft

Horst Hutter

Nietzsche and the Modern Crisis of the Self

Nietzsche, when writing in 1888 that his name would one day be associated with "something tremendous—a crisis without equal on earth, the most profound collision of conscience . . ."[1] seems neither to have exaggerated his future fame and notoriety, nor to have misunderstood our modern predicaments in any way. No other thinker, in my opinion, helps us more in defining the many incongruities and contradictions of our self-understanding. These involve particularly our ability to manage the powers that we have called forth but to which we are becoming enslaved. We have created technology as the instrument of our will to power by which we aspire to rule the planet as its "owners and operators." We know this as globalization. Yet in the process of actualizing this willing, we have succeeded also in transforming ourselves into the servants of this giant machinery of production and control. Seemingly our will to power has been transformed surreptitiously by our own very actions into a will of powers beyond our control.

Sharing in the paradox of thus being both supremely powerful and abjectly powerless, we have lost all previous certainties and faiths. The very command to "know thyself" that stands at the beginning of this development has dissolved into the recognition that both the knower and the known of this injunction seem to be equally illusory. Not knowing who we are, we know even less what we should do. Indeed, not only is the goal lacking, but even the traveler as well as the road suspiciously resemble the fond projections of thirsty desert runners.

Yet Nietzsche not only diagnosed the modern spiritual condition, he also suggested ways and techniques of overcoming by which we might resubstantiate ourselves, redefine our goal, and travel on firmer roads. In the following I shall try to answer the question of what one should do on the grounds of his rather pessimistic but seemingly "true" teaching. What are the modes of self-care and self-cultivation by which, taking inspiration perhaps from the example of Baron von Münchhausen and his swamp, we might extricate ourselves from our self-incurred predicaments? Not knowing who we are, what are the tools that Nietzsche suggests we use to actualize our ignorance so as to become who we are?

Latter-Day Socratics and the "Counter Tyranny of Reason"

The virtues of reason as the ruling "part" of the soul are more affirmed in Nietzsche's denial of these virtues than in most celebrations of them. A "great reason" seems to be silently at work in Nietzsche's decentering of the self: it is the indirect regent of his many attacks on traditional notions of "soul," "reason," the "passions," the "ego," in short, all conceptions of unity and harmony obtaining in the multiplicity of souls. At one time it suggests a discourse of nay-saying under the sharpness of which inherited aggregations of the soul disassemble and wither. Uncovering secret lies, poses, and inauthentic postures, it shakes the structures of the soul to their foundations and exposes their hidden designs. Henceforth—this "great reason" warns—these structures can only be maintained by a dishonest hiding of the fissure or by the despicable contentment of small creatures able to live in ruins.

At other times, however, the "great reason" appears as "a mighty ruler," an unknown sage that steers the spirit by yes and no saying to a new order for the agonistic multiplicity of souls. Here a new nobility, a kind of warrior aristocracy of selves seems to design itself; a conception of "reason" richer, more ambiguous and infinitely more dangerous than its predecessors.

The oscillation between yes and no in the descriptions of reason given by Nietzsche is closely linked to the contemporary disintegration of reason, to the increasing tensions that appear in the oppositions in both modern politics and, more importantly, the inner politics of modern humans. The tensions, hidden conflicts, and silent struggles seem to occur between these two psychic poles: on one side a mass, a "mob," a "herd" addicted to the stupefactions of virtual worlds and secretly ruled by the inner demons of small envies, petty rancors, and slowly poisoning jealousies. The loss of metaphysical hopes have loosened the traditional "bonds" of the soul, seriously destabilizing the soul's habitual and customary regime.

The resurfacing from within of the mob of hitherto hidden agencies of the soul, of new tyrants and new kinds of tyrants, only dimly suspected *as* tyrants, is followed by the disarray and disintegration among the old "ruling" elements. "Reason" there is in headlong flight; caught in internecine bickering and mutual deconstructions, it loses the substance it once seemed to have. There it "rules" still, but merely through and in its externalized form, namely the giant and standardizing "third worlds" of economy and virtuality. The movement seems to be from appearances of a pain-pleasure continuum to a " reality," and on to its shortest shadow, the "beyond" of the virtuals.

The rule of psychic virtuals is duplicated in the political world by new kinds of tyrannies, be they commissary and serial dictatorships—military juntas—clever banks aggregating "monies" as enabling solvents, small national or polit councils, nocturnal or diurnal, all dedicated to "security"—great helmsmen—saviors of the people, as well as all kinds of providentially guided leaders and orderers, movers and shakers. Political herds and masses are ruled by a double strategy of enlisting the new psychic tyrants slumbering in them for both the works of manipulation and of repression . The stupefaction and the disuniting of souls into virtual hazes are here combined with secret mobilizations of anxieties through guilt and terror: television plus death squads and a deliberate cultivation of the need to punish.

Nietzsche's soul experienced the terror of dissolution fully, as it began its long slide from the Apollonian certainties of its Christian regime into its disintegration and its dismemberment as a fitting disciple of Dionysos. This movement from Apollo to Dionysos, with its desperate attacks on unity and simultaneous cries for help precisely from the disappearing center, its joyful celebration of psychic polycentricity, and its unitary affirmation of multiplicity induced a brightness of vision and a gift of prophecy rarely equaled since ancient times. His autodiagnosis enabled him to sketch the outlines of a new "harmony" but it was not sufficiently strong to resolve the inherent contradictions arising from his impossible life-tasks. How to save the great "reason" by attacking reason?

Perpetrator and victim at the same time, the song of his loving soul barely contained the disharmonies of his "great longing" and his "great contempt." Yet out of this impossible, nearly "divine" tension emerged the arrow of a new longing, a new stretching of the bow and the attempt thereby to forge a new psychic regime.

The ambiguity of Nietzsche's attack on "reason," his attempt to befriend it as his most intimate enemy, is mirrored in the manifold contradictions of his teaching. One such contradiction is his lovingly unfair attack on Socrates and Plato, and the elaboration of this attack into an attack on the continuation of the psychic regimes formed by them and their presumed successor Jesus. His positive teachings seem to cohere as little on the ontological level as they do on the existential level. They are all suffused by the same tension and irreconcilable oppositions between incompatible but necessary affirmations. In my opinion, they cannot be resolved on any static level of either ontology or existence. But if they are regarded as forward-tending forces in the soul, resultant in a new form of self-cultivation, their contradictions may be entrusted to the guidance of our "great" reason which, following Nietzsche, seems to be the "body."

This trust—certainly a species of *amor fati* and justified only by the aims and intentions of the arrow to a new eternity—may not be sufficient; yet it is certainly an arrow loosed from Nietzsche's arc. The following may be seen as an attempt to trace the trajectory of this arrow. Whether the arrow reaches its aim is not entirely in the disposition of archers, unless archers are able to unite their "little reason" with the great reason moving through them and with the archer.

My proposal hence is not to analyze the contradictions of Nietzsche's teaching as such, be they metaphysical, ontological, or existential. Ultimately perhaps, the contradictions between eternal recurrence, will-to-power, and the *Übermensch* are irreconcilable. As a comprehensive teaching, however, Nietzsche's does not seem to be different in this respect from all other teachings, scientific, philosophical, or religious. I shall focus, instead, on the ambiguities in Nietzsche's attitude toward "reason" alluded to above. These seem to point directly toward a reexamination of the dimensions of soulcare and self-shaping, once so much a part of philosophy but until recently so neglected by philosophers. The question to be asked is, how the tensions in Nietzsche may be applied practically to the development of a set of techniques useful for some individuals for shaping themselves in the direction of a new self-overcoming.

All willing requires a goal and a rank order of "goods" arranged as way stations on the journey to this goal. It requires consideration of the material conditions and the obstacles one is likely to encounter along the way. The goal for Nietzsche is the development of a new species of human beings that ultimately will merge into the creation of the *Übermensch*. The present generation of humans, including the free spirits that may arise, hence are preparatory and transitional. Preparation and transition require efforts of willing and self-shaping. They require the creation of techniques for self-creation and the elaboration and cultivation of new virtues. And they require initial acts of destruction.

When seen from this perspective, the contradictions within Nietzsche's comprehensive teaching are merely one particular formulation of the general tension that rends the human soul as well as the mind as such. This tension has always led to life strivings in the direction of unity and harmonizations. But perfect "identity" has never yet been achieved; unity has always only been the goal of willing and never its existential end point. One novelty of Nietzsche may consist in the fact that he no longer seems to posit unity as a goal of striving, but looks toward a way of living well with multiplicity, temporality, and polycentricity. Human reason thereby becomes reintegrated into a larger movement— a movement of a "great reason" perhaps—within which individual reason is contained. Human fragility, temporality, and the understanding of reality in terms of a "principle of insufficient reason" are thereby placed into the fore. We do not begin from eternity, identity, and unity but rather from their opposites, which we are unable to deny, think away, or engineer away. As such we are at the mercy of forces that work through us, that present us with ineluctable "givens" which we do not own and do not control. (This is our situation within global technology.) But we may to some extent shape these forces in the creativities of our interpretations. For if all interpretations of the world are human creations, then the elaboration of such interpretations may be the only access to freedom that we have.

The Problems of Freedom and Responsibility

One of our most deeply rooted convictions seems to be the opinion that we are endowed with a free will that enables us to choose between good and evil. Establishments of criminal justice are based on this assumption and thus hold us responsible for our actions by the infliction of punishments. Notions of retribution, restitution, and deterrence, associated in various combinations with systems of punishment, all presuppose that there are individual doers who are free to choose acts that shape their futures and amend their pasts. While notions of retribution and restitution seem clearly oriented to the segment of the flow of time called the past, deterrence is, rather, oriented to the future segment of time, in the direction of which we only seem to be able to hope and will. Past and future are linked by the fiction of responsibility in which individual wills both will backward in retribution and restitution and will forward in the fear and anxiety induced by the threat of deterrence. Punishment as such seems to be a social and political mechanism for the shaping of wills.

Systems of individual responsibility, inculcated via punishments, appear to be rather recent in the development of civilizations. The logical paradox associated with their founding, so clearly visible in Plato's *Laws*, has always been the fact that they wish to create something they first must presuppose, namely, individual responsibility.[2] We may thus see them as attempts by the great founders of past orders of the soul, as cultivating devices intended to create individual freedom and to do so by shaping the multiple forces of the soul through a perhaps noble and salutary myth. This myth functions by imposing a direction on the manifold strivings in individuals. It is accepted—an acceptance induced by punishments—as an "as if" narrative that shapes the attitudes taken by individuals toward their own inner conflicts. In the conflict between desires and prohibitions it functions as the third and reconciling force that tilts the balance in favor of prohibitions. As such it imposes an interpretation of past actions by willing to change the effects of those actions for the present and the future. It is thus a fictitious "willing backward" so as to change potential willings forward.

Nietzsche attacks this myth head-on and destroys its metaphysical, psychological, and existential plausibility. To him there is no free will, the will is not a psychological faculty independent from reason or even associated with reason; and responsibility is exposed as a lie that merely has the conviction of millennia on its side as a supposed "truth." The firmness and matter of course manner in which an opinion is believed—however many people may believe it—does not establish the truth of an opinion. Neither does fondness for an opinion, nor its authority.

But Nietzsche goes further. Pointing to the manifest fact that the deterrence theory of punishment seems to be one of the best-refuted opinions, he not only disputes the "truth" of individual responsibility but also disputes its nobility. It may once have been a salutary and noble political fiction, but it may no longer be so considered. The passage of time and the evolution of modern science have laid bare the ignoble basis of this once "perhaps" noble fiction. Retribution in particular, but also restitution, point to a particular psychological propensity naturally given in all humans, the desire and need to punish when suffering has been experienced. The origin of systems of punishment hence cannot be seen to lie in the wish to implement a presently given "free will" by actualizing its future freedom; that is merely its justifying fiction. Rather, it may be seen as an attempt by noble philosophical souls of the past to provide an outlet for an individual's need to inflict suffering on others upon the experience of one's own, inevitable suffering. By providing such an outlet a salutary political goal might be achieved, namely, the containment of cycles of vengeance. The human need to inflict suffering on others when suffering is endured has the natural political tendency of drawing ever-wider circles, involving and engulfing ever-larger numbers of human beings, even whole societies in general crises of sacrificial cruelty. Systems of punishment thus deliberately continue the ancient practices associated with all religions, to use measured doses of punishments at regular intervals of a social rhythm of life by way of sacrifices and by public punishment of scapegoats. The little and controlled "cruelties" of punishments are thereby pitted against the eventuality of large and uncontrolled "cruelties." It is a case of defeating Big Vengeance by using little vengeance against it, a case of playing cruelty against cruelty and thus the "devil against Beelzebub." And as such it already implies a germ of human autonomy by creating the shadow of an "individual" in the psychic possibility of a management of "punishments," by opposing a present desire for vengeance in reinterpreting the past, and thereby to achieve a freedom of the future from a repetition of the past. The "permission" given to acts of "little cruelty" necessarily implies self-cruelty in the inhibition of immediate vengeance. And this self-cruelty seems to have in turn needed as *its* justifying theory, the invention of systems of metaphysical punishment with the interpretative creation of fictitious entities to whom one can leave one's desire for vengeance. At least the divinity present in the monotheistic religions seems always to have asked believers to leave their vengeance to the divinity.

The ignoble foundations of the noble fictions of punishment and responsibility have been laid bare for us by the events of modern history. These bespeak not a decrease of human sacrifice but an enormous increase of cruelty and sacrificial crises. No previous epoch seems to have known wars of such a ferocity, such a cruelty, and such a generality of destructiveness as has our epoch. In view of such occurrences Nietzsche insists not only that the once noble fiction must be now seen as ignoble, but that it itself is now at the origin of the widening of the cycles of vengeance. The only counterpoise to this widening of vengeance is either an extinction of humanity in an act of self-destruction, or the growing actuality of the permanent psychic deformation of the culture of the last man, or a destruction of the old myth of the soul and the creation of a new one. Nietzsche also foresees and plans for both latter eventualities.

The task of the new myth of the soul is nothing less than to free humanity from the cycle of vengeance, to liberate the future from being chained to a repetition of a past. This chain is now simply the control of the present by systems of generalized vengeance and systems of punishment. The breaking of this chain requires first a freeing of the present toward an open future, by way of a reinterpretation of the past in the form of a new "noble" myth. The past is to be redeemed from the curse of continuation by way of its total acceptance in either *amor fati* on the level of the individual, or *eternal recurrence* on the political level. Both doctrines are to be new devices of self-cultivation, enabling souls to transport their need to punish into the past, rather than as hitherto practiced, into the future.

This new way also involves the employment of cruelty. Aspects of this cruelty are the renunciation of all hitherto accepted comforting illusions, a total acceptance of present suffering and the radical impermanence of everything, and a refusal to respond to suffering by the infliction of suffering. And these are primarily acts of self-cruelty, which may represent a total spiritualization of cruelty as a future possibility. They hence propose a new achievement of autonomy by way of a new ordering of the forces of the soul. Passions, reason, and the virtues have to be redefined. Such redefinitions may permit a new way of leading the inner war, guided by a new vision, a new employment of the negativities of the soul. In this case too, freedom would be coextensive with the autonomy gained by the whole of an individual in playing his "parts," negative or positive, against each other. But the whole would no longer think of itself as a permanent I or immortal soul in a mortal body, and would not even aim at such eternal identity. Nor would it think of itself as a permanent ego endowed with a "free will."

Self-Cultivation among Latter-Day Socratics

Given that Nietzsche both denies individual freedom but also affirms at least the possibility of a new vision of autonomy and freedom, a legitimate question to be asked is still the old question, first seemingly proposed by the Stoics, namely, what within me and what without is within my power and what is not within my power. The Stoics appeared to have quietly presupposed in merely raising this as a possible question, that psychic events at least are able to be divided into the two classes of "within my control" and "not within my control." Moreover, the Stoics presuppose that there is an I that can raise this question and make such a decision. But Nietzsche denies all such notions of the "transcendental" and undermines the logical, metaphysical, and ontological foundations of it and of the very possibility of this question; yet he also assumes that management of psychic forces is both possible and, above all, necessary.

For why otherwise would he have written, and written in such a way as precisely to initiate processes of self-transformation and new modes of soul management? The Stoic question would hence have to be rephrased, a rephrasing that would take into account the fact that there is no self, no permanent I, no ego, and that only fictions have hitherto functioned as the guarantors of their existence. The question would more properly become one that asks what is given as distinct from what is not yet given but may become so. And the fulcrum for human autonomy would be the grey area between the "not yet given" and the "may become." And Nietzsche assumes that my reinterpretation of the "given" can tilt the process in the direction of one or other of the possibilities within what "may become."

The deconstruction of the traditional concepts of "soul," "ego," "body," "will" require a fundamental revisioning not only of these former entities as "parts" of a total human, but also a reconfiguration of human totalities as such. And this revisioning and this reconfiguration must occur without any assumption of a unitary agent doing them, nor any belief in free will, nor yet any divine intervention. Everything is process, and the passage of time has only relative "forms" of duration: all things are merely temporary shadows on the surface of a primal and ever-present chaos.

So to the first question: what, according to Nietzsche, is given, or more precisely, what is given to someone observing his or her inner processes (assuming that we are able to distinguish an "inside" from an "outside")? It is to be noted that in the very raising of this question, a remnant of "transcendentalism" is already presupposed. The question of what the manner of this "transcendentalism" is may be put in abeyance for the moment. Now it appears that for Nietzsche virtually everything is given. The "givens" include not only my entire "genetic" inheritance but also my cultural shapings, including all my past decisions regarding my interpretations and their residues in me, and including my whole "Christian" or other configuration of soul. The whole past, understood both phylogenetically and ontogenetically, with all of its joys and sufferings is thus given. My "soul" is merely a term for the multiplicity of drives, affects, wishes, thoughts, and intentions that have ever occurred in my interiority. Timewise these extend into the infinite past and comprise all vegetal, animal, and human strivings that have ever occurred to produce me, this very short duration in time.

Moreover, I "am" at present both a war and a peace, a great reason, a plurality with one sense, a herd and a shepherd. And the name for the totality of these manifestations is "body." Soul is merely the name of something about my body, and my little reason, that is, my I and my spirit are merely tools and playthings of my "self." The body is governed by an overall intentionality, a "self," which is a powerful commander and an unknown sage. It dwells in the body, it *is* the body. The body and its great reason does not merely say "I," it does "I." Thought and reason, as traditionally conceived, are deceptions that hide the fact that they are at all times under the control of something larger and more powerful. This illusion is created by the fact that the "selves" of many bodies are in a downward intention, they wish to go under, because they believe they lack the ability to "create beyond themselves." Hence, they despise life and the body and are angry at life. Their cunning gaze hides an "unknown envy" that seeks revenge against life.

It seems evident from such a proposal that everything is given to me since I did not create nor ever govern the entity called "my" body. But my "self" has the ability to create beyond itself, unless my "unknown envy" denies this ability. The only two possibilities for a future for me are then either to create beyond myself or work on and with my envy: either creation or destruction or perhaps both. Insofar as Nietzsche's configuration of the human totality is not merely a description of what is but is also an injunction to act differently and to intend the future differently, the question must be asked who so understands the totality and who may intend differently toward the future, either as creation or as a destruction working on "unconscious envy." Two different kinds of selves seem to be required here, a self that has hope of creation and a self that has contempt for life. The question of Nietzsche's transcendentalism imposes itself at this point.

Reconfiguring Human Beings

Let us recollect here some aspects of Nietzsche's teaching: the denial of unity, the emphasis on war, the recognition of self-cruelty at all levels of human culture; the unity of love and contempt, of "good" and "evil," the denial of "being" (with its concomitant denial of "nothing"); the denial of causality and of "doers"; the recognition of multiplicity and the insistence on primary chaos; but also the radical fissure of time at the "present," separating past from future that is the centre of time, both weightless and the most weighty; both eternally passing and the *nunc stans*. And finally, the difference of the future from the past. Who can join them? Who speaks all these things?

The duality of that which dwells in the body, that which both is and is in the body, namely the self and the evolution of selves into ascendant and decadent selves, frequently dwelling in the "same" body, nevertheless would seem to require a duality of treatments. The self uses its tools to compare, overpower, conquer, and destroy but it does so in two modes, creating beyond itself and aiming at its own extinction.

The self, itself a duality, is both one with the body and different from it: the human totality at any rate seems to consist of four "parts": the self, the body, the entities (wisely) commanded and used by the self, and all three together. But then it is not clear whether or not the "best wisdom" of the body is the same as the self or merely an instrument of the self that is not identical with the body, which always contains more reason than (the) best wisdom.

What then is the matter with this duality of the self, "the mighty ruler" and "unknown sage"? Why does it sometimes believe in its ability to create beyond itself and then why does it not? The attitude recommended by Nietzsche here is one of creation and destruction, either singly in the case of unambiguous, non-schizoid selves; or combined into the joining of the attitudes of great longing with great contempt in the case of a schizoid self.

But who would "do" these two things and even make these choices? Who would "do" differently in the three modes of thought, deed, and image, given that the "wheel of the ground does not roll between them"? Is it a body, a self, a great reason, or all of them together?

The question of who "decides" either to create beyond himself and who despairs of his ability to do so, can perhaps be answered, if we examine more closely the schema for the interior arrangement of a human totality proposed by Nietzsche. According to this schema the human totality contains a "trinity" that comprises as the "father," as it were, the body, as the "son" and "ruler" the self, which is both the same and not the same as the body; and finally, as the "holy ghost," the tools and toys of the self: thoughts, feelings, the ego, etc. (For convenience I shall refer to the tools collectively as the ego). The chain of command clearly seems to be one where power moves from body to self to ego, with the self being the executive organ of the body that merely employs the ego and prompts its concepts and suggests its feelings of pleasure and pain. It is also clear that the self has the crucial position of being the commander who clearly governs the ego but whose authority in regard to the body is ambiguous.

We may assume that Nietzsche believes that human beings have always, or at least for a very long time, been such totalities. If this is so, then it becomes puzzling how the traditional regime of the soul, proposed by Christians, could ever have been established in such a being. For according to the Christian schema, the "soul" is a unity, comprising thoughts, feelings, and sensations, that rules or ought to rule the "body." The proper relationship of soul to body is one where the soul exercises tyrannical rule over the body and where the body is demoted to a mere expendable tool, in any case a tool that best be used for as little time as possible. The preaching of contempt for the "body," the mortification of the "flesh," and the denial of pleasure and the sometimes active seeking of pain are all instrumentalities by which the rule of the "soul" over the body is to be established and increased. How could the true ruler of a human, namely the self, ever have been fooled into accepting such a complete illusion as the Christian regime of the soul seems to be? Unless by such a regime, the power of the self, its seeking, comparing, conquering, overpowering, and destroying, is enhanced. The self's turning against itself, implied in the Christian regime of the soul, its contradictory and schizoid impulses, both to create and to turn against itself in self-destruction, have then led to the development of a particularly strong and high and mighty ego, which rules in usurped majesty over the whole. Being the eyes and ears, its spirited thinking and dialectical analyzing have endowed it with the perspectival illusion, where it believes to govern while all the time being secretly governed. Hence the edifices of dogmatic philosophies and dogmatic sciences with their illusions of omnipotence, an omnipotence enviously modeled on the nature of the divine maker. And the ego has succeeded in forgetting its real position in suppressing its awareness to the point where even the self is divided in its consciousness and has, as it were, forgotten *its* destructive side. This complex of the ego, with its far-ranging thoughts, strictly controlled feelings, deadened senses, and selective memories, has then become crystallized into a quasi-permanent mask of which the bearer has forgotten that it is a mask and as such merely part of a game of theater. It would not be unreasonable to designate this congealed mask of an ego with the word *personality*, an acquisition of centuries of Christian training in self negation. The personality would then stand in dialectical opposition to the self, perhaps fruitfully designated as the "essence" of the body, in an order of "counterstriving conjunction." One is tempted to call this whole complex of the received ego, the "personality" with its illusionary "free will," an ignoble lie in the soul that has become accepted as the truth, a "bad" faith that presents itself as a "good" faith. And this bad faith, almost the foundation of modernity, has now become life-threatening, not only for its bearers but also for others, by the power of its marvelous instrument of technology. Technology, in turn, can be characterized as the congealed lie of the power ego.

Nietzsche was illuminated by the duality in his nature, as both a Christian decadent and a new beginning, into a powerful recognition of the dangers and the urgencies of the present culture. Fully facing up to the "terror of the situation," he attempted to forge a new regime for the soul, a new psychagog. This task, which he set himself, required a double strategy, involving both destruction and creation, and an employment of both nay-saying and yes-saying. To begin with, what would need to be destroyed, as is evident from his reconfiguration of the human totality stated above, is what I have called "personality." Personality, the result of a particular series of employments of the ego by the self, needs to be deprived of its comforting illusion of potency, of its belief that it is in charge, and that it can freely dispose of all the forces at its "command." Personality needs to be disassembled into the series of now unconsciously worn masks of essence that it really is; the masks need to be recognized as masks, that is, they need to be understood as devices that conceal and reveal at the same time. This understanding would then reawaken the consciousness of the self as the true ruler.

Secondly, then, the historically acquired tendency of the self to turn in upon itself in its activity of destruction, its "death instinct," needs to be affirmed away into the past. By contrast, the other activities of the self, namely, listening and seeking, comparing, overpowering, and conquering, also turned inward together with destruction, need to be affirmed by being repeated into the future. Eventually the self might then acquire the ability of a life-affirming play in which it freely employs the masks available to it as "toys and tools." Such play would necessarily also include all the masks of "slavishness," that is, those masks that the self developed in originally turning in upon itself. All the past would thereby be "redeemed" into a new order of the soul. And the acquisition in the past of the death instinct would thereby have performed its important labor in producing and enhancing the depth of the human animal. For it seems clearly to have been "slave morality" that has produced the greatness of human interiority.

This new development would seem to proceed in four stages, two of them destructive and two constructive. First a destruction of the illusions of the power ego; secondly a dissolution of the fictitious "substantiality" of the ego, which would simultaneously also destroy the "death instinct." Thirdly a reestablishment of the proper and seemingly more "natural" rank order and command structure to again obtain between body, self, and ego; and fourthly the attainment after the stage has been cleared, as it were, of a new free play of masks of the self.

I have above spoken of things such as self, body, and ego, as if they were self-subsistent entities, with something like an actor called self wearing masks. Such a perspective would entail a return to traditional, metaphysical conceptions. Lest the impression is gained that I therewith surreptitiously reintroduce into Nietzsche's thinking concepts that he denies, I hasten to add that these terms do not refer really to entities but to processes of war and peace. Following Nietzsche one would have to deny all substantial entities. And even though body, self, and totality may be only aggregations of wills-to-power, constantly changing with a void as the "subject" of change, they nevertheless present themselves to any introspection as if they were "given" as fixed entities, certainly an illusion on Nietzschean grounds. We may think of them thus not as stable identities but as movements of similarity through time, having at least the structure of similitudes. They partake of the structure of those life-affirming and "truthful" illusions that constitute the horizon of a sustainable life. Perhaps a more appropriate way of thinking of them would be as repetitions by means of which the past repeats itself via the present into the future. Each one of these repetitions, never really the same but only similar, would be a repetition of aggregations of wills-to power. Each aggregation would be both a war and a peace. The "great noon" of the recognition of these as repetitions of similitudes would then enable the self that is one aggregate of such repetition, to affirm its power over all repetitions of the past and the future. Such affirmation, a truly "unbounded yes," would impose upon the eternal recurrence of *similitudes* the great breeding device, the idea of the eternal recurrence of the *same,* which is an idea that is both destructive and creative.

The application of the idea of eternal recurrence of the same would thus be an attempt to "impress the stamp of being upon becoming." It would function as a selection tool by which "bad" repetitions, that is to say, life-denying ones, could be affirmed away into the past, and life-affirming ones would be affirmed into the future. The self would "destroy" life-denials and death instincts by confining them to the past. It would create beyond itself by repeating life-affirmations into the future that are as much as possible identical and the "same."

Would this, however, not reintroduce the old duality of good and evil and place the self again into an attitude of at least partial negation? To a certain extent this is so; but if we think that the negations inherited from the past, that is, life denials, are built into the fabric of the modern self, the negation of destructive repetitions would have to be preceded by a total affirmation of the self as it has become, as it is now in the present. Such an affirmation would also be an affirmation of all self-destructive repetitions. The only total affirmation of a death instinct would be the attainment of its aim, namely, death and extinction. Hence, Nietzsche's frequent reiteration that the idea of eternal recurrence, if it were to get hold of souls, would destroy many weak human beings. Hence also Zarathustra's advice to the despisers of the body to "say farewell to their bodies—and thus become silent." Hence also Zarathustra's exhortation to the judges of the pale criminal, who has achieved sublimity in condemning himself, not to be lenient but to kill him for "there is no redemption for one who suffers so of himself, except a quick death." And we might also say that upon the doctrine of will-to-power, life-denial certainly also is a form of power, but it is a lesser power than life-affirmation, which is hence preferable merely on grounds of greater power: it might thus become the new "good" as against the old "evil" of life-denial.

I shall conclude the above remarks by again affirming that my purpose here is not a critical exploration of Nietzsche's teachings as ontology, metaphysics, or cosmology. Thus, my remarks on eternal recurrence should be seen only in their aspect of revealing Nietzsche's teaching of a new mode of self-cultivation. The specific techniques of this self-cultivation now need to be explored, after having described the general framework within which they may be placed.

Soulwork

The pattern of a self-cultivation has to begin from a recognition of the nature of the garden that needs to be cultivated. That is to say, we must grasp the basic outlines and design of the culturally acquired residues and traces left in the original soil by prior efforts of cultivation, and this in regard to both the traces left by an individual's ancestry, taken in the widest possible sense of the term, as well as those left by one's own attempts at transforming one's given inheritance. It is tempting to gain such an understanding by constructing an ideal model of human development, as it is meant to be in accordance with a more or less fixed delineation of "human nature" as it once was enacted, and as it has since become distorted and deformed, but to which it may again be restored. Thus, it is easy to think of the "natural" rank order "body-self-ego" that I have developed above as something that once characterized a more genuine and primitively original form of self-development. As a corollary of such a view, one might then think of Nietzsche's effort at reeducating humans as one that wishes to undo the deformations wrought by "slave morality" and thus to restore "master morality." This interpretation of Nietzsche, a very frequent one, then sees the restoration of master morality to consist in the overthrow of slave-like restraints on aggressivity and on the desire to dominate; it recommends a quasi fascist theory of desublimation.

Quite apart from the fact that the subjects of a master morality, as seen by Nietzsche, do not at all fit the picture of the future higher selves developed in Zarathustra, such an "essentialist" interpretation would seem to be totally opposed to Nietzsche's view that the soul has a history and that we cannot go back to prior stages. We moderns are composites containing elements of both master and slave ancestors. We are storage houses for the accumulated wisdom of our ancestors living in and through us, including the "slave" inheritance. From slave morality we derive a much greater intelligence and cunning, cleverness, emotional refinement, and general depth of soul than the ancient subjects of master morality ever had. One should remember Nietzsche's insistence on the historically established total victory of slave morality and the threat that this victory poses for the human future. We must thus start from the recognition that we are "slavish" to the same extent that Nietzsche himself was, with this slavishness being an unremovable overlay over prior layers of the evolutions of the soul. Our "primary" nature is thus a historical succession of accretions of previous "secondary" natures.

What Nietzsche proposes is the creation of a new secondary nature that replaces and overcomes that which we now believe to be our second nature but which is merely the latest stage and top layer of our primary nature. Such willing forward is to retain the positive and transform the negative elements of slavishness. Indeed, this project of self-transformation must use the ego instruments developed by the slavish self, such as introspection, self-criticism, rational linear, as well as dialectical, thinking, deception, ressentiment, irony, and "unknown envy," that is, all the "negative" qualities of slaves. It must utilize the very countertyranny of reason supposedly developed by Socrates to contain his decadent instincts, which Nietzsche sees as the origin of the "slave revolt in morals" (of course, together with the Hebrews' transvaluation of values). In short, it must accept its starting point as Socratics endowed with the wisdom of Odysseus.

Returning now to the proposed "proper" rank order between body-self-ego, we are constrained to pay particular attention to the divided nature of the modern self, for it would seem be a defining characteristic of the slave's self to be divided against itself in self-love and self-hatred. It is a self in which death-instinct and life-instinct are combined in the same movement of the spirit. Some further aspects of this division are the mixtures of honesty with dishonesty, of anxiety with repressed rage, innocent envy and "guilty" guilt, and of course hypocrisy, the favorite virtue of all kinds of "morality screamers," who so frequently become prominent in the political world. On the positive side of this division we might list industriousness, attention to detail, perseverance and great intelligence without much understanding, versatility, motleyness and jarring disharmonies in thoughts, feelings, and bodily awareness.

Now to a closer look at how we might conceive on Nietzschean grounds the division in the self. We may think of the modern spirit, an instrument with the "ears" with which the self listens, as caught in a double bind: in the spirit a great longing is combined with a great contempt, and every effort of (self-) creation is also an effort of (self-) destruction. When the ego says "I," but the body does "I," and when such doing is also an undoing, then we face a complex of "personality" in which ego and self are so intertwined that an ego illusion, originally produced by the self, is maintained by the ego as a "truth" even against the continuous production of other, perhaps more fruitful illusions. The ego has hardened into a complex of fixed modes of thinking, feeling, and doing that merely appear to itself as a form of autonomy because they are automatisms. Such automatisms are maintained by the very energy of creation/destruction siphoned off from the body, but with the continuous repression of the awareness of destruction into the unconscious, into "unknown envy." The self has created for itself an ego and has endowed it with tyrannical power that can be maintained only by a lack of awareness of what it is doing. The Socratic countertyranny of reason has quite simply become the tyranny of "reason." The pull on the energy of the instinctual economy of the body must be increased, so as to maintain the double labor of projection and repression. This very bad use of always limited energy then also induces the typical deformations of personality, as given in intellectual overdevelopment combined with emotional underdevelopment, both housed in a sickly body. Truly a "bundle of wild snakes and a heap of diseases." An "ideal" decadent would thus be someone who has fixed, dogmatic principles, a sharp and cunning intellect, primitive emotions, and a neglected, machine-like body. Such a person with "firm character" and an indomitable "free will" would seem to be a modern norm. But in reality such a person does not think, *it* thinks in him, nor does he feel, but *it* feels, and the whole is done with a phenomenal lack of self-awareness. Everything is automatic, and willing is locked into a series of such automatisms. In such a condition, human beings do not live, they are being lived, and the world is perceived in a distorted manner by which the judgments of others about oneself are taken as one's own judgments. The individual lives in the general "one-does-one does not," always anxiously intending the other but always serving the ego in a narrow way that both inhibits awareness of one's true interests and maintains the "selfishness" of the ego. Pity in such a state comes to be seen as altruism, which in general then is merely a form of underhanded egoism. Thinking only of himself, an individual in this condition merely implements a general play of masks automatically prepared for him.

The ego complex of automatic thinking, automatic low feeling, and awkward moving, the whole called free will, is so encrusted in the self that the ego becomes the self's illusion of mastery. Awareness of the whole is maintained only dimly in the unconscious. The inversion of the body/ self-ego relation as one of ruler-ruled into its opposite, is maintained in the false consciousness where the ego thinks it is the ruler and the self and the body are being ruled. The whole, I suggest, should be termed "personality" with the underlying self-structure of the body being the "essence" of a human being.

Now, personality and essence are in perpetual conflict, whereby personality constantly wins pyrrhic victories over essence. The first step in a creative self-transformation would hence have to be the becoming aware both of one's personality and one's own essence and their mutually destructive war and their co-dependency. This war is being fought by means of the mechanisms of anxiety and guilt. Both of these destructive emotions, already consuming more energy than is required for living, then take even more energy to maintain the thought structures that are needed by ego to feed anxiety and guilt. If we think of anxiety as the dread of punishment at an indefinite time in the future, but still in this life, we may think of guilt as the dread of punishment in the "life beyond" into which one could enter at any moment. The dogmatic metaphysical systems that have been erected so as to anchor anxiety and guilt in some absolutism "beyond" are established and maintained with an enormous expenditure of psychic and political energy robbed from the self's reservoir of energy left for self-awareness

Nietzsche posits the virtue of honesty, perhaps a modern variant of Socratic *parrhesia*, as the virtue of access to unraveling the nefarious personality structures of the modern self. But how can someone who is honestly dishonest or dishonestly honest ever achieve that moment of impartial self-awareness in which it becomes absolutely clear what is going on? Even if attained in flashes of insight in moments of great danger or in periods of great threats to survival, as in illness, how can this flash of awareness be maintained and structured into a continuous presence of mind to oneself? A continuous act of mental attention to oneself that does not judge and does *not* condemn? How is such self-remembering to be achieved? For changes can only be made after a sufficient time of self-awareness, whereby one's completely particular ego-self configuration becomes known. What can be done and what should be done only reveal themselves at the point of insight into one's own particular deformation and pathology.

Honesty would have to be slowly acquired in a cultivation of the virtue of some mindedness, Nietzsche's virtue of virtues, "*Besonnenheit,*" in long periods of solitude, self-observation, the remembering of dreams and daily lapses, and the discernment of patterns in the automatically proceeding functions of thinking, feeling, and doing. And most importantly, an honest showing of oneself in the therapeutic conversation to a significant other, a dropping of the strong tendency to hide oneself. Only this way can one's "honesty" be made to remain honest, as it were.

In ancient philosophic schools, such as Epicurus's Garden, or the Academy, or later in Christian monasticism, such self-observation could be felicitously combined with the revelation of oneself to an other in some confessional structure. Self-revelation could be practised without anxiety or the constant desire to please—one was among friends, after all—honesty with others might be forcefully encouraged and dishonesty in self-revelation be punished. One would enter within such a confessional structure for a while into a disciple-master relationship in which the "master" is trusted to "know" more than the disciple; at least, he sees what the disciple cannot see. Such loyal following by the disciple of a master would have to last for as long a time as the disciple needed to achieve a sufficient degree of sustained self-honesty and impartial self-awareness. Only at this point could the inner tyranny of the various automatisms of personality begin to be broken and the labor of reassembling the self-ego begun. Throughout all of this the disciple had to continuously subordinate his ego to the ego of the master and to "present his soul" as well as his mental attention to him.

The practices of reeducation in ancient schools most likely also included a reeducation of the "body." Since the energy for the tyranny of inner automatisms is always taken from the body, these automatisms are also established in the body in patterns of automatic movements and postures. These bodily automatisms would reinforce the ego-self automatisms in such a way that a thought structure would be borne by a conflictual emotional structure, in turn grounded in physical mechanisms. In principle the work of dissolution of these automatisms could start either at the top end, the thought structure—at the middle, in the emotions of the spirit—or at the bottom, in the physical habits; or it could begin by working simultaneously on all three levels. The aim would be the breaking of all automatisms of associations between thoughts, feelings, and postures. But in any case, these automatisms always also had to be destroyed in the body. We do not have sufficient evidence regarding the kinds of body work employed in ancient schools, but they likely included breathing exercises, reeducative gymnastics, and in general forms of body work similar to the martial arts or to yoga. Which kind of body work would be required would naturally depend on the specific deformation in question and on the type of self to be produced by philosophical reeducation, which was either contemplative or active. But in all cases the spurious rhythms of automatic existence would cede to rhythms of a deeper, more genuine existence.

Now, Nietzsche did not experience such a school, he never even found an appropriate setting, nor did he ever, as far as I know, found such a school. Perhaps he did intend to create such an entreprise out of some of his friendships. But he never succeeded. He was all alone and he had to do it all by himself. And it is amazing, given this paucity of opportunities, how far he actually got. Indeed, the paucity of experience in this regard may even have inspired him to conceive of new modes of self-cultivation and a new form of friendship related to self-cultivation. His utterly solitary existence, and the deep suffering from his solitude, almost to be called "divine" in their grandeur, made him reach that profounder layer of the human soul, a layer wholly beyond any individual ego or any human willing. From this abyss he could see the unmitigated suffering of a world, seen only as will-to-power, almost as if by an "invitation." The vision then enabled him to specify and to communicate the nature of modern culture, its life-threatening propensities and the thoughts and practices that would be required for its reconfiguration.

Techniques of Self-Overcoming

The second stage in self-cultivation is reached when self-honesty has developed into the inner presence of a guiding light or a voice by an effortless and continuous holding of the mind to itself. Then might arise the understanding of how to begin to destroy the mechanisms by which the illusion of a firm ego identity is maintained. These appear chiefly to be identifying and considering.

Nietzsche's vision is above all useful at this stage of self-transformation. The techniques he conceptualized and practiced show their utility here in the efficacy of destroying the mechanisms of identifying and considering. I believe I can discern in Nietzsche's life and writings specifically five different such techniques. These five techniques seem to be solitude, friendship and the dialectic between solitude and friendship, writing, extreme care in selecting the three kinds of "food" required for the well-functioning of a human being, and, fifthly, withdrawal from actual politics as well as from the marketplace, a modern form of the Epicurean "*lath biosas.*" Before discussing these five techniques and the three kinds of food, however, it might be useful to clarify my notions of identifying and considering.

In identifying, the ego accepts the *reaction* of its self to the *action* of others as wholly belonging to him, no matter how contradictory they may be and how little they may serve the development of "what is truly one's own." The attitudes of others, their benevolence or malevolence, are accepted as one's own and are integrated into a general attitude of constant and anxious self-examination. The self thereby also accepts moods and opinions of others as its own and "forgets" its real needs, which emerge from within. The boundary between self and other is never firmly established and the emotional bond with others in which we originally arise is never wholly broken. No independent development of one's essence in terms of its own needs is ever fully achieved. Living in fear of the opinions of others, the ego-self here does things, not because they might be good for oneself, but because they please others. Concern for reputation and for appearances, status seeking, and the tendency to live in the eyes of others are the chief characteristics of such a personality. Everything hurts, and only the approval of others brings any kind of comfort. Yet others are also always envied and resented. Self-overcoming here takes the form not of a growing into one's task in terms of one's given potentialities but of a false acceptance of what others believe to be the task for oneself. The individual lives under the tyranny of oughts and ought nots and rarely achieves independence of willing. The herd mentality completely dominates.

Moreover, the ego has no distance from the succession of thoughts, feelings, and wishes proceeding always automatically within itself. It is wholly identified with this succession and is unable to take a distance by confining each moment to its temporality in saying: "Yes, this too is transitory, but it is not I but only a temporary reaction of mine." The ego is thus unable to construct out of the inner processes of self a personality, or rather a series of masks appropriate to the real needs of the self at any given time

Lacking the ability to resist any tendency within itself, unable to say no to a stimulus, a wish, or a desire, such a personality easily succumbs to any one of the compulsions and obsessions made available to itself by a culture. The inability to transcend herd consciousness then frequently takes the form of addictions and subterranean erotic compulsions.

The "cure" for such a condition and the beginning of development of one's essence in terms of its own good, would consist in acquiring the ability to say no to one's "instincts" and impulses, that is, to choose to identify only with those of one's inner tendencies that are judged to serve the "true" interests of one's self. And these need not be the same all the time. Hence, the ability of a self to construct for itself several "personalities" and several ego identities in a lifetime, that is, the ability to wear different "masks." A self unable to do this is someone for whom "altruism" may serve as a mask of confused self-interest. It can also be seen that the mechanism of guilt is originally created out of identifyings, maintains them, and in turn is maintained by them. It can further be seen how the entire morality of pity and altruism, so fiercely attacked by

Nietzsche, is a political tool for maintaining selves in a condition of immature identifying and in a lack of real self-development. Selves are thus maintained as herd automata. The overcoming of inner automatisms of identifying would permit the abolition of the false dichotomy egoism-altruism, the refusal to accept the "goods" of others as one's own "goods" and the restoration of a proper selfishness in which the self seeks its own (good). The definition of the good would vary from self to self and from period to period in the self's own development in terms of its pre-given possibilities. As a cautionary addition I should like to assert here that such a true "selfishness" is entirely compatible with the search for a "common" good, commonly pursued by strong individuals, but it is only fully compatible with any public good in the "good" society, which, however, does not exist anywhere. The only way in which individuals have been able to construct "good societies" hitherto is in the communities of noble friendships that they have succeeded in establishing. The "good" society existed for Nietzsche only in such friendships. The political and economic "goods" pursued in the societies in which we are obliged to live may only be partially compatible with the goods of reeducated autonomous selves. That is to say, the herd goods now sought, such as "equality," comfort, and wealth, can function as the mere basis that is necessary for the developments of a new "greatness of soul," a new nobility. Finally, it needs to be remarked that all strivings of all selves, including the consciously reformed strivings of autonomous selves, would in any case and at all times remain the actions of that "entity" that works through our bodies and our selves, using the ego as its tools and toys, namely the "entity" that Nietzsche calls will-to-power.

The second psychic mechanism for the maintenance of ego illusions is a considering in which the self puts its own interests ahead of the interests of others with which it interacts. It does not see and cannot place itself imaginatively into the perspective of others so as to see the impact of its actions on others. It thus remains caught in the self-seeking that in Christian morality is opposed to altruism. In dealings with others it considers only how the advantage of self may be realized; but this advantage is defined so narrowly as to never include a larger perspective, which ultimately would be more propitious for its self-enhancement. It cannot forget itself and thus cannot see how its own ego illusions entrap its actions in a self-defeating spiral. Its self-concern thereby serves to maintain its self-division in an awareness without awareness. As such, considering is the obverse of identifying and continues the domination of self by the anonymous "other" that has originally produced its schizoid condition. It thus is constrained by its own ego-structure to repeat into the future the similitude of enslavements inherited from the past.

The "cure" suggested by Nietzsche might be for the self to forget itself so as to find itself in an awakening to the "true" dimension of a higher selfishness. This would obtain once it no longer looks within and no longer identifies its totality with its automatisms, but looks without and achieves its enhancement in subduing and conquering its environment. Thereby it would remove the blinders of vision that it imposes on itself by its own "egoism."

The beginning of Nietzsche's destructive, no-saying labor of self-transformation would consist in a ceasing of "egoistic" actions by the self in a letting-be of itself and in a total self-acceptance as it has become. The clarity of seeing and hearing resulting from such self-acceptance cannot be attained, however, in a situation of interaction with other schizoid selves that constantly reinforces one's own self-division. Conforming to a society that is dominated by slave values merely reinforces the domination of the slave's ego in oneself. This participation in the general emotional structure of one's society can only be stopped by a withdrawal from society into solitude. Solitude tends to produce self-acceptance and an understanding of one's inner automatisms and limitations. It awakens awareness of the existential possibilities that are realistically available. The self-contradictions of the schizoid self involve a turning around of vital energy in a circle-like cage, maintained by the need of the limited ego to identify and consider when interacting with other schizoid selves. Solitude removes the need for identifying and considering and permits a return to one's "higher" self hitherto hidden under the mask of "altruism." It creates this "higher" self.

But the insights and greater self-awareness attained in solitude need to be tested in the company of others, yet not the company of others caught in the slave-mentality. Hence the need for the search for friendship with like-minded individuals, either equal to oneself or more aware than oneself. Both solitude and friendship with like-minded individuals had been provided in the setting of ancient philosophical schools. This setting was not available to Nietzsche, but solitude and friendship were. Hence, Nietzsche constructed a technique consisting in a dialectic between retreating into solitude and then reemerging into friendship with others. Solitude permits the self to become aware of its own unknown envy for the fullness of life elsewhere. And friendship permits the self to transform its envy and its desire for vengeance against its own life into a virtuous envy. In a proper friendship with an equal or with someone superior, self can use the negativity of its now-conscious envy in a striving for self-overcoming, motivated by the desire either to surpass an equal friend or to become like a superior friend. In friendship, narrow selfishness can be overcome and the voice of conscience, hitherto buried deep in "essence," awakened. This voice of conscience is a "knowing" together with a friend of the "true" dimensions of one's existence and its limitations. Mimetic rivalry between friends repatriates thus envy into the desire of the self to overcome itself. This desire to overcome, it will be remembered, had been locked in the schizoid self in a split, comprising the desire for greater strength and the desire for one's own demise. Mimetic rivalry between friends now permits the self to understand that its previous ego-illusion had been only a mask. It understands the mask *as* a mask, and begins to realize that it is free to adopt other masks. Moreover, in such friendship, both selves get the opportunity to don different masks and try them out in other games. Friendship forces such engagement in other games, provided it is animated by mimetic rivalry.

Nietzsche proposes a model of "true" friendship, quite unlike the Aristotelian model, which had included the notion that such "true" friendship can only occur between two "good" persons. This already presupposes a commonly held notion of goodness, a goodness already more or less obtained by persons entering into a "true" friendship. But Nietzsche cannot rely on any accepted notion of goodness, precisely because modern society does not offer one. Rather, he is in the precarious cultural position in which the search for the human good has been forgotten and abandoned and the "way" has been lost. The "way" or the "ways" have to found again in a rekindling of the desire to search for one's own good. Nietzsche proposes a new model of friendship that can rekindle this desire by way of mimetic rivalry. "Genuine" friendship in this fashion does not obtain between persons already good and knowing this, but between persons who know that they do not know their own good and that they hence need to search for it, with the definition of what such a good might be being entirely left open. Such persons are, to begin with, related to themselves by both friendship *and* enmity; self-friendship, always being the basis for friendship with others, thus contains in the case of schizoid selves always also self-enmity. Nietzsche's model of friendship contains this self-enmity with its corollary of "enmity" for the other, actualized in the desire to surpass the other. Friendship thus is structured by the dialectic of enmity. And this dialectic of enmity fuels the search for the true good in an enlightened *"egoisme à deux"* by way of clear attention to oneself.

The integration of enmity into friendship, always silently required in any friendship, is consciously advocated by Nietzsche. This then requires the introduction of a *pathos of distance,* presupposed by the link between self-friendship with self-enmity. *The pathos of distance* exists already within each (schizoid) self. It needs to be externalized in mimetic friendships by an alternation of closeness *and* distance between friends.

This closeness/distance and its pathos is marvelously described by Nietzsche in the peregrinations of Zarathustra. Zarathustra ascends and reascends to his mountaintop; he also continuously returns to his friends. On the mountain Zarathustra renews his wisdom and takes cognizance of the masks he has donned when in company. He devises new masks and re-descends to his friends so as to perform new games with them. He suffers from the distance from his friends when solitary, and he also suffers when with his friends from the insufficiency of all friendship. He gains the insight that the *pathos of distance* is the necessary starting point for schizoid selves and a permanent feature of self-overcoming. Self-friendship can never be total for human selves, because it always contains "enmity" for the self-egos to be left behind and abandoned. Likewise, friendship will always contain "enmity" in the form of mimetic rivalry. The quest for the "good" is unending and self-overcoming has no fore-seeable upper limit in a condition of permanent "self-identity." It re-quires contempt for the baseness and stupidity in oneself and others; a baseness to be redeemed by being accepted into the starting point of one's self-overcoming. Love and longing are hence inextricably linked with contempt. Only with contempt can self-transformation be directed into self-overcoming. And the awareness of this contempt can only be maintained in mimetic friendships. Friendship thus permits a continued openness of the self to the manifold possibilities of new masks, new modes of self-creation, and new friendships. In this vision, every friend-ship is both a home and a prison, and every house of selfhood something to be escaped from. No rest, as they say, for the wicked.

Nietzsche actualized in and through his writings a third technique of self-transformation. They served the destructive purpose of unmasking the various modalities of the modern death instinct and the ego illusions on which they are based. Hence, his mode of writing as a war, constantly in the posture of attack, never in retreat. Simultaneously, writings have always permitted the maintenance of newly acquired self-understandings by externalizing the inner impartial observer. They served this purpose also in Nietzsche's quest. But in his case they could do so only by having his writings continuously pointing to himself, to his own "ipsissimosity." Hence, necessarily everything he wrote for publication is autobiographi-cal. Any other "scholarly" writing could not have served his purposes. And since the trajectory of his and our self-overcoming is not finished, his writings are also necessarily incomplete and do not contain doctrines as such. Rather, they are invitations to his readers to engage in their own quests for self-transformations. He invites his readers to forget him, not to follow him, and to find themselves by forgetting him. Each genuine way of transformation is structured by one's inheritance and is hence

totally unique. It can only begin in the discovery of one's "givens."
Nietzsche's "doctrines" hence are mere temptations and attempts to be
surpassed by future such attempts in a continuous quest aiming at the
creation of a higher type of human being.

Nietzsche's correspondence, unlikely to be intended by him for
publication, served him to maintain the dialectic between closeness and
distance in his own friendships. They are records of mimetic rivalry and
enact the *pathos of distance* by providing closeness at a distance. They
thereby also permitted the incorporation of his new selves into most of his
old friendships, mirroring his own growth and, unfortunately for him, not
mirroring to him the equal growth of his friends in most cases. Most of
his friendships seem either to have developed into coldness at a distance
or seem to have remained in the solitude of a one-sided virtue of gift
giving. At best they provided loving acceptance without sharing and understanding, a fact that is perhaps not unrelated to his final breakdown. He
truly had to do it all by himself, a feat reflecting in its impossibility the feat
of Münchhausen in extricating himself from a swamp by his whiskers. But
his suffering in this regard may have become fruitful for us in providing
the insight for us into our own necessities for self-struggle. His very failure
may thus be the condition for the success of his followers.

Nietzsche was throughout his life a selective and careful eater and
a very selective and careful reader. Moreover, he spent a great deal of his
time on his lonely walks outdoors, believing, it seems, that good ideas
come only with the drawing of fresh and clean air. We may think of
everything we take into our totality as a "food" for different parts of
ourselves. Thus, gross material things to eat and drink are food for the
"body," air is food for the self and its thinking and feeling, and sensory
intake, including the mediated intake through reading and listening to
music is "food for thought," that is to say, for the creation and destruction of ego-masks of the self. "Good" food on all three levels is necessary
for the well-functioning of self and the structure of its self-overcoming.
But what is "good" food?

Nietzsche believes that no one recipe for good nutrition suitable for
all selves is possible or desirable. Each individual self must decide for *itself*,
precisely how *it* needs to nourish itself so as to achieve its maximum
strength. There are, however, some individual rules that a self may discover
within, upon impartial self-observation and with the ability to say no to
"spurious" desires, that is to say, those desires the satisfaction of which
merely feeds one's weakness and lack of autonomy. "Good eating" would
hence require the ability to choose and to avoid what one takes in.

Everything taken in, what is eaten, drunk, read, heard and listened
to, and what is breathed in, needs to be metabolized, that is to say "
compared, overpowered, conquered" and if need be, "destroyed." Proper

foods, taken in proper amounts at appropriate times maintain the totality and permit it to grow. To be sure, the body stops growing after a while, but there are no real limits for the growth of the self, except perhaps the imaginary limit of the *Übermensch*. Hence, the criterion of choice, open in regard to specific contents, is limited in regard to the ability to assimilate, integrate, and eliminate. One of the prime foci of attention for any future philosophy, according to Nietzsche, would lie in a careful study of the question of nutrition and the proper practices of eating. The possibilities for cultural renewal arise from a consideration of this most important question. This is particularly evident in regard to the "foods" taken in through reading. The worst possible practice here would be the practice now followed in the academic and intellectual world: the reading of vast amounts of low-level stuff, increasingly lower materials consumed at an increasingly rapid rhythm, not understood, not integrated into the self, indeed even unrelated to genuine self-concerns. The consequences would seem to be increasing spiritual and mental indigestion, bloated intellectual entrails and the regurgitation and excretion of a vast amount of culture barf at ever lower levels of re-digestibility. And most of this "reading" moreover is done in bad air, bad lighting, and is accompanied by the consumption of large amounts of increasingly worsening material foods eaten at ever more rapid rates. The whole culture seems to be in an ever-faster race to ego domination and self-extinction as witnessed by the declining ability to read and write good things. Hence, Nietzsche does not tire to emphasize the importance of slow and careful reading, and the setting of deliberate limitations to the amount and nature of things read. (I wonder what he would have said about "foods," had he witnessed the contemporary trends to physical obesity and cultural confusions.) Books are initially strangers that have to be integrated into the discourse of self-friendship. Some fit and others only fit badly the actual level of self-development. Some increase the range and depth of spiritual vision, others decrease it. In any case, everything read has to be made "one's own" before it can be said to be understood.

Quite similar considerations would apply to conversations with others. Some are fruitful and mutually enhancing, others are not. Some persons are to be sought out, others avoided. Mutual understanding is always difficult to achieve and presupposes a whole range of similarities. In a society characterized by great diversity of types, genuine meetings are rare and misunderstandings almost the norm. Here too, friendship is crucial by defining the limits of mutual intelligibility.

The state, according to Nietzsche is an institution dominated by the life-denying values of the market place. It in turn spreads the illusion of the power ego and attempts to universalize it in imposing slave-values on peoples or in reinforcing them. Both the state and the marketplace are extensions of the power wills of herd automats, whose wills, it may be remembered, combine life and death wishes. Successful participation in either hence requires that one cultivate those personality formations that insight into one's essence—one's "secret" self—would oblige one to deny. Genuine self-cultivation is incompatible with "political" activity.

Yet Nietzsche's teaching is profoundly political, but it is political in the sense of aiming to shape the human future. Meanwhile, in an age of preparatory humans, self-cultivation safely can occur only in the interstices left by wealth and political power. In that the planet still contains enough emptiness for the expansion of great-souledness, the "free spirits" that arise will not continue to enslave themselves to the deadly games of wealth and power. But direct contact with the ruled is also impossible; they are entirely caught in the automatisms of the last man. Hence, the only viable politics today for Nietzsche, is the great politics of human self-overcoming. The stages of this refashioning detailed by Nietzsche need not detain us here, for they must be settled by the efforts at self-cultivation of our descendants. For now, the path indicated is one of the life removed from the state and from the marketplace. In this preparatory age, human reason can only be redirected toward "ideal" and future possibilities in the orientation of one's strivings. The god that may come can only dance on a stage of the human soul made ready for his appearance.

Toward a Newly Dancing Socrates

If properly handled, the application of the five techniques described above should result in a sustained awareness of processes of inner automatisms of thinking, feeling, and moving. The illusions of firmness of ego-identity and free will, the mistaken belief that the ego is in charge and the destructive control fixations of the ego will have been dissolved. The resultant "emptiness" with the ability to suspend the mechanisms of identifying and considering, then should enable the self, the human essence, again to reconstruct the dance of reason.

This dance of reason had once been exemplified by the movements of its grand master Socrates. For it does not seem exaggerated to consider the Socratic *elenchus* as an appearance on the stage of the human soul of a dionysiac *logos orchoumenos*. The unfairness of Nietzsche's attack on Socrates is particularly evident when we consider how masterful a dancer Socrates was. The depth of Nietzsche's critique, however, may be appreciated if we think of it as directed not so much against the manifold personalities of this ancient pied piper of the soul but against their distortions and degenerations in the various types of modern Socratics (of which Nietzsche was one).

Reason is captured and distorted in the various forms of power-ego illusions, a kind of "consciousness" residing in dogmatic philosophers, dogmatic scientists, and dogmatic priests. In Nietzsche's words, the preachers of sleep or death and the flies and showmen of the marketplace of ideas, as well as the cold-blooded managers of state idolatry with their front-men, the morality screamers, are all united in the endeavor to increase the sleep of members of the herd and to prevent the many sleepwalkers, including themselves, from ever waking up. These too are Socratics, who have even managed to turn the openness of the method of the modern empirical sciences, a marvelous resuscitation of the Socratic *elenchus,* into closed fixtures. They have transformed a critical instrument into a panzer ego that moves forward with increasingly machinelike and destructive precision. Now, Nietzsche radically destroyed the foundations of these complexes of the modern soul in himself and in all those who wish to follow. His critique of Socratism seems entirely justified. Starting from these "givens" we may now briefly trace the steps that Nietzsche enacted in his endeavor to reanimate the Socratic dance of reason. It may be helpful in this tracing to recall the image of a bodily dancing Socrates presented in Xenophon's *Symposium* (II, 1–24).

Upon observing the performance of the troupe of the Syracusan dance master, a by then roughly forty-seven-year-old Socrates expresses his admiration for the graceful movements of one of the dancers who in the harmony of his movements appeared even more beautiful than he had appeared at rest (II, 15). Socrates then states his desire to take lessons from the dance master so as to learn how to move his body in a simultaneous symmetry of all his parts, for in this manner a person must dance who intends to increase the suppleness of his body. This suppleness is attained not by the one-sided gymnastics of long-distance runners or prizefighters. Rather, the moving of all of one's parts harmoniously together will increase health and the enjoyment of the simple pleasures of eating and sleeping, besides enabling a person to reduce an outsized paunch. And if age and ugliness should prevent one from finding a suitable dance partner, such harmonious gymnastic may be performed alone. Thereupon Socrates confesses that just the other day Charmides had caught him dancing alone and had believed at first that Socrates was mad, but when Socrates described to him his intention of achieving a harmony of movements, Charmides had begun his own solitary practice of shadowboxing.

Dancing involves the suspension of the ego: the less the conscious ego is involved, the greater are the strength and suppleness of movements. In dance the Dionysiac forces surface from the depths of the body's great reason, ego awareness is perfectly focused and relaxed, and the self steers the totality of parts in a free play of forces. But this free play of forces is the result of a long discipline and a great restraint of forces, a restraint no longer exercised by the conscious ego. The autonomy thus gained is a freedom of the movements of the self that has used the constraints of the ego in a ceaseless process of self-shaping. The rigidities of the power ego have been utilized for the creations of discipline and are then dissolved in the dance.

Nietzsche proposes the image of the dance as a metaphor for the reshaping of the rational ego in a sequence of learning how to think, learning how speak, and learning again how to write.[3] Learning how to see, the basis of thinking, involves not reacting to stimuli, not willing, letting things come as they will, postponing judgment, and accustoming the eye to calmness and patience. With these tools—seemingly, acquisitions of a long training under the domination of slave moralities—one does not permit all doors of the soul to remain open, and one gains control of all the inhibiting and excluding instincts. Automatic compulsions are then broken in a slow, mistrustful, and recalcitrant posture of consciousness. The ingrained slave habits of objectivity and of being servile before every "fact" are overcome by being turned upon themselves, as it were. The self learns to suspend identification and to consider things in themselves.

Learning how to think begins from the recognition that, initially at least, when looking within, one discovers that *it* thinks, *it* feels, and *it* moves. These *its* have to be taught how to dance, so that the stiff clumsiness of modern spiritual gestures and their bungling graspings can be transformed into a dance of concepts. This dance of concepts would seem to be precisely a reconfiguration of the Socratic *elenchus,* so much characterized by ironic distance and the withholding of assent.

Nietzsche has little to say in regard to learning how to speak but he certainly had a whole lot to say in regard to learning how to write. And he not only preached here, but he practiced a form of writing that makes the mind of his readers spin with dizziness. This dizziness is not overcome unless one has learned to follow the dance of one's own soul.

Notes

1. Friedrich Nietzsche. *Ecce Homo.* "Why I am a destiny" in Karl Schlechta (ed.) *Friedrich Nietzsche. Werke in drei Baenden.* (Muenche: Carl Hanser Verlag, 1966) Vol. 2, p. 1152.

2. cf. Plato. *The Laws of Plato.* Thomas Pangle (trans.) (Chicago: University of Chicago Press, 1980) 853b–863a, pp. 245–57.

3. cf. *Twilight of the Idols.* In ibid., VIII, 6–7, p. 988.

12

Globalization, Technology, and the Authority of Philosophy

Charlotte Thomas

I

English-speaking readers of Jacques Ellul's *The Technological Society*[1] miss the playfulness of its original French title: *La Technique: L'enjeu du siècle—Technique: The Gamble of the Century*. It is also easy for contemporary English-speaking readers to miss, in the wash of daily dispelled predictions about the beginning of the twenty-first century, the last section of the last chapter of *The Technological Society*, "A Look at the Year 2000." In "A Look at the Year 2000," Ellul does not so much make his own predictions about our contemporary society as criticize the predictions of some of the prominent intellectuals of his day; but Ellul's critique is illuminating. Ellul is consistent and emphatic in his insistence that he makes no recommendations about how best to respond to the technological determinations of society that he catalogues, so we should not make the mistake of looking to him for answers to current questions about human prospect and philosophical possibility. It is perfectly within the scope of his work, however, to appeal to the technological society for greater clarity about the way things are; that is, to hedge our bets in the technological gamble.

Let me take a moment to clarify my understanding of Ellul's stance on the issue of predictions, solutions, and advice. One must, I think, pay special attention to Ellul's explicit intentions if one is to avoid falling into the mistake of reading *The Technological Society* either as a self-help book for technological despair, on the one hand; or, as a fatalistic account of no-exit technological determinism, on the other. In Ellul's 1964 Foreword to the Revised American Edition he articulates his position very clearly:

> As to the rigorous determinism [which might lead to pessimistic interpretations of this work], I should explain that I have tried to perform a work of sociological reflection, involving analysis of large groups of people and of major trends, but not of individual actions. I do not deny the existence of individual action or of some inner sphere of freedom. I merely hold that these are not discernible at the most general level of analysis, and that the individual's acts or ideas do not here and now exert any influence on social, political, or economic mechanisms. . . . To me the sociological does not consist a collective sociological reality, which is independent of the individual. As I see it, individual decisions are always made within the framework of this sociological reality, itself preexistent and more or less determinative. I have simply endeavored to describe technique as a sociological reality.[2]

For Ellul, the technological society is not the sum of the individual actions of its citizens, but rather a reality that is independent of individuals and determinate of individual action. So, Ellul makes no attempt to account for the possibility of individuals freely acting against or outside of the sociological boundaries he describes. Such acts are possible, but not probable or predictable. The order Ellul describes is, thus, not fatalistic. Each individual, although determined by the technological society, is not, in principle, exhausted by it.

Before one becomes too sanguine about the human capacity to transcend the determinations of the technological society, Ellul jumps in with three examples of external factors that could change the course of history; that is, three things that would not only signal divergence from the determinations of the technological society, but could actually alter those determinations. According to Ellul, the three external factors are: (1) the postapocalyptic vision—a war with few survivors and a radically different set of survival issues, (2) enlightenment—a critical mass of people see the spiritual price of technology and lead the march for a life with different priorities, and (3) *Deus ex machina*—God intervenes to change the course of history or the nature of humanity.[3]

It would be easy to take this list as a joke—Ellul presents the three ways that his account might avoid fatalism, and they're all absurd; but I think Ellul means all three of them seriously. Even the one that reads most preposterously, divine intervention, is the subject of much of Ellul's theological writing. On the other hand, Ellul makes it clear that there is no good reason to think any of the three ways out of the technological society will materialize. An overwhelming probability indicates that society is and will be determined by the conditions described in the book. We can think about whether or not the technological gamble is a good bet, but it is too late to take our chips off the table. Ellul turns the argument of his foreword around one last time before closing, when he rises to a sort of altar call for those who might assert their freedom against the necessity of the technological society's determinations:

> In the modern world, the most dangerous form of determinism is the technological phenomenon. It is not a question of getting rid of it, but, by an act of freedom of transcending it. How is this to be done? I do not yet know. That is why this book is an appeal to the individual's sense of responsibility. The first step in the quest, the first act of freedom, is to become aware of the necessity. The very fact that men can see, measure, and analyze the determinisms that press on him means that he can face them and, by so doing, act as a free man. If man were to say: "These are not necessities; I am free because of technique or despite technique," this would prove that he is totally determined. However, by grasping the real nature of the technological phenomenon, and the extent to which it is robbing him of freedom, he confronts the blind mechanisms as a conscious being.[4]

The technological society confronts us with necessary determinations on human life and we, as humans, are specifically determined such that we are able to assert ourselves against its necessities. This paradoxical view of human nature sets the stage for all of Ellul's sociological observations. As dismal and inevitable as Ellul paints Western culture's technological decay to be, human freedom always hovers in the margins as the means of unlikely transcendence. Ellul wrote *The Technological Society* for an expressed purpose: "to arouse the reader to an awareness of technological necessity and what it means."[5] Ellul, the anti-activist, ends his foreword to the American edition with a description of his work as " a call to the sleeper to awake."[6]

It is impossible to argue that technology has done anything but grow in influence since the publication in 1954 of *The Technological*

Society. And while many might have argued with Ellul in the '50s about the ubiquitousness of technological determinations on human life, few would do so today. Our apparent technological self-consciousness is not the wakefulness Ellul hoped against hope for, though. Instead, we embrace technology precisely in the way Ellul described as indicating a complete abdication of our freedom to technique's determinations.

This is indeed the future Ellul foretells in the last chapter of *The Technological Society.* Ellul describes the late-twentieth-century human race as "beginning confusedly to understand at last that it is living in a new and unfamiliar universe."[7] Technology emerged and was embraced with a definite end in mind: insulation from and mastery of nature, Descartes's old goal. But, instead of floating above nature with technological immunity, humanity subordinated itself more and more to technology's promise, until the promise didn't matter anymore and technique emerged as its own justification and reward.

In the final section of this last chapter, "A Look at the Year 2000," Ellul casts a hard eye on a feature published in 1960 by the Parisian weekly magazine, *L'Express.* The article consisted of excerpts from works by prominent scientists on their predictions about society in the year 2000:

> As long as such visions were purely a literary concern of science-fiction writers and sensational journalists, it was possible to smile at them. Now we have like works from Nobel Prize winners, members of the Academy of Sciences of Moscow, and other scientific notables whose qualifications are beyond dispute. The visions of these gentlemen put science fiction in the shade. By the year 2000, voyages to the moon will be commonplace; so will inhabited artificial satellites. All food will be completely synthetic. The world's population will have increased fourfold but will have been stabilized. Sea water and ordinary rocks will yield all the necessary metals. Disease, as well as famine, will have been eliminated; and there will be universal hygienic inspection and control. The problems of energy production will have been completely resolved. Serious scientists, it must be repeated, are the source of these predictions, which hitherto were found only in philosophic utopias.[8]

Ellul's flabbergasted inventory of the absurdities propounded in *L'Express* continues:

> Knowledge will be accumulated in "electronic banks" and transmitted directly to the human nervous system by means of coded electronic messages. . . . There will no longer be any

need of reading or learning mountains of useless information; everything will be received and registered according to the needs of the moment. . . . [N]atural reproduction will be forbidden . . . [and] artificial insemination will be employed to conceive in vitro the children of parents who embody the masculine and feminine ideal—preferably parents who have been dead long enough to enable objective judgment about the merits of their lives and work.[9]

Well aware of the power of technique to accomplish the inconceivable, Ellul focuses his criticism of these absurd visions of the year 2000 not on the reasons why it might be unreasonable to believe society might evolve in those ways; but, rather, on the enormous personal and spiritual cost such a society would command:

Consider, for example, the problems of automation, which will become acute in a very short time. How, socially, politically, morally, and humanly, shall we contrive to get there? How are the prodigious economic problems, for example, of unemployment, to be solved? . . . [H]ow shall we force humanity to refrain from begetting children naturally? . . . How shall man be persuaded to accept a radical transformation of his traditional modes of nutrition: How and where shall we relocate a billion and a half persons who today make their livings from agriculture and who, in the promised ultra rapid conversion of the next forty years, will become completely useless as cultivators of the soil? How shall we distribute such numbers of people equably over the surface of the earth, particularly if he promised fourfold increase in population materializes . . . etc.[10]

According to Ellul, there is only one possible answer to all of these "hows," "a world-wide totalitarian dictatorship which will allow technique its full scope and at the same time resolve the concomitant difficulties." [11] All of this is absurd, of course. Since the year 1984 arrived in non-Orwellian fashion, it seems that every day brings another prediction unfulfilled. Yet Ellul's final jab at the scientists quoted in *L'Express* and at contemporary scientists in general, suggests a vaguer but more believable threat:

We are forced to conclude that our scientists are incapable of any but the emptiest platitudes when they stray from their specialties. It makes one think back on the collection of mediocrities accumulated by Einstein when he spoke of God, the state, peace, and the meaning of life. It is clear that Einstein, extraordinary mathematical genius that he was, was no Pascal; he knew nothing of political or human reality, or, in fact, anything at all outside his mathematical reach. The banality of Einstein's remarks in matters outside his specialty is as astonishing as his genius within it. It seems as though the specialized application of all one's faculties in a particular area inhibits the consideration of things in general. Even J. Robert Oppenheimer, who seems receptive to a general culture, is not outside this judgment. His political and social declarations, for example, scarcely go beyond the level of those of the man in the street. And the opinions of the scientists quoted by *L'Express* are not even on the level of Einstein or Oppenheimer. Their pomposities, in fact, do not rise to the level of the average. They are vague generalities inherited from the nineteenth century, and the fact that they represent the furthest limits of thought of our scientific worthies must be symptomatic of arrested development or of a mental block. Particularly disquieting is the gap between the enormous power they wield and their critical ability. To wield power well entails a certain faculty of criticism, discrimination, judgment, and option. It is impossible to have confidence in men who apparently lack these faculties. Yet it is apparently our fate to be facing a "golden age" in the power of sorcerers who are totally blind to the meaning of the human adventure.[12]

There are none more respected in our society than scientists, and in many ways, their status is deserved. Even those of us who see the dark side of Western culture's seemingly unstoppable advance enjoy living in the contemporary world, with all of its successes: transportation, communication, medicine, etc. But for those who do not have enough experience with science to appreciate the narrowness, repetition, and specificity of most scientific work, labs are labyrinths of knowledge and scientists are heroes of untangling the mysteries of the cosmos. As heroes who know how to negotiate the labyrinth, scientists seem to have access to truths we mere mortals cannot approach, and their opinions on more global, cultural, or humanistic topics are met with an undeserved deference. As Ellul saw, such deference is deeply misguided and very dangerous.

II

I was listening to an NPR talk show not long ago that featured a tête-à-tête between William Pollack and Christina Hoff Sommers. Pollack, author of *Real Boys*,[13] presented and defended his thesis that American culture stereotypes boys, pushes them to be tough and independent, and alienates them from the more sensitive, interactional, and dependent aspects of their nature. Sommers, author of *The War Against Boys: How Misguided Feminism Is Harming Our Young Men*,[14] argued against Pollack, claiming that boys were doing fine in American culture, at least as fine as the rest of us. She proposed that the crisis that Pollack described in the life of American boys was an attempt to jump on the bandwagon of diagnosing victimization, and that the argument of *Real Boys* was a revision of an early '90s push in the women's movement to criticize the ways girls were treated in schools and brought up to envision themselves. Girls had had their day in the sun of victimization, and Pollack stepped up to give boys their chance.

Sommers talked eloquently of the histrionics of victimization. She soberly argued against identity politics. She acknowledged the truth of broad claims about the difficulties of emergent adolescent sexuality, but staunchly denied reports of an epidemic of girls' suicides or an increase of boys' crimes of rage. She had done her homework and could account for the misleading results and overblown conclusions of the studies that were quoted to substantiate Pollack's arguments; but all of the callers and the mediator of the program showed an unmistakable preference for Pollack's side of things.

Some of this, of course, comes from the love of victimization, which seems to be the coin of the realm. No one seemed to want to confront the individual responsibility implied by Sommers's levelheaded dismissal of Pollack's announcement of imminent disaster. But much of the difference in attitude, I'm convinced, stemmed from Sommers's status as a philosopher versus Pollack's credentials as a psychologist. Many of the callers would begin their comments by thanking Pollack for his research. Pollack used the first person ceaselessly when referring to the data derived from his modest series of interviews with boys. The commentator once went so far as to interrupt Sommers and throw the ball back to Pollack by saying, "Well, Professor Pollack, since you've actually done one of the studies in question, what can you tell us about this issue?"

It did not matter that Sommers was more articulate, better read, more reasonable, and less apocalyptic in her account of things. It did not matter that she could quote Pollack's work more fluidly that he could.

Pollack had done studies. He was the scientist. He had found the golden bough and returned to tell his tale.

This common and disturbing mistake, the fallacy of scientific authority, leads us to the less present, but deeper question of what constitutes real authority. Technology and globalization seem key boundaries to the contemporary context of this age-old question. The easily learned, powerful, and broad application of technology has eroded (or, at least transformed) the tropes through which we make sense of our experience. The sage is today mistakenly identified with the scientist, as Ellul points out, and the wisdom of the sage has been replaced with specific technical knowledge.

Globalization is an extensive complement to technology's intensive transformation of the idea of authority. There is no longer a mountaintop to which one may climb for enlightenment—all of them have been demystified through aerial photography and extreme sports. Nor is there thought to be an ancient culture living in a remote jungle which is privy to secrets of the ages. Anthropologists, whom we must trust because of their status as scientist-sages, have made it quite clear that cultural differences are ubiquitous but not hierarchical. There is no more truth to be found in any one given place than in any other; except, of course, the sterile lab, which can be set up anywhere.

As technology makes things easier, it also conditions our desires toward that which it can facilitate. We see the power of technology both because it is powerful and because it seduces us into desiring those things over which it can have power. Likewise, the broadening of our perceived geographic boundaries through communication and transportation advancements opens up possibilities for seeing things more globally; but such possibilities also lure us into thinking in terms that are amenable to the view from thirty thousand feet. Our contemporary situation is rife with temptations and incentives to look at the world either through a microscope or a telescope, but rarely simply with our own eyes.

Thus, technology and globalization move us toward a view of the world very much like one of the ones described in Meletus's, Anytus's, and Lycon's famous accusations against Socrates in Plato's *Apology*—an inquiry into things in the heavens and below the earth. Like the rest of the charges against Socrates, at least as Plato tells the story, the accusation could not have been farther from the truth. When he first addresses the jury in the *Apology* and again when Meletus takes the stand, Socrates insists that he has never been engaged in inquiries concerning things in the heavens and below the earth, but has committed himself exclusively to conversations about human matters. With a few significant exceptions, the dialogues support Socrates' description of his endeavors. This concern with human life and the things that can be made sense of as a result of immersing oneself in it is what Socrates most fully embodies and what the contemporary world leads us farther and farther away from.

It is the nature of technology not only to change the way we see ourselves and the world, but also to change the nature of our desires and capacities. So, it is worth thinking about whether the technological society not only makes it less attractive for us to shape up a life of philosophical inquiry into the ethical and political dimensions of the human condition, but whether it also may transform our thinking such that the conditions for the possibility of political philosophy are shakier. It seems likely that humanistic philosophy, and every other form of inquiry that does not yield to the Cartesian equation of truth and certainty, will seem less and less plausible as science redefines authority for our culture.

Philosophical authority requires an attunement to reliable reference points for truth. Socrates' allegory of the true pilot from Book VI of Plato's *Republic* illustrates this sense of authority. Socrates describes a ship owner, sailors, and a stargazer on a ship. The ship owner is taller and stronger than anyone else on the ship, but he is a little deaf, his eyesight isn't great, and his knowledge of navigation is on par with his sight and hearing. The sailors are described as an unruly mob engaged in a constant struggle to take the helm. None of the sailors has any knowledge of navigation, and they deny that navigation is an art that can be taught. Each sailor claims that his experience is the only relevant credential, and his ability to force his way into power is the only relevant concern. So the sailors are always dogging the shipowner, sometimes tricking him into giving the helm over to them or detaining him forcefully while they take the ship on wild, drunken excursions. The sailors have no idea that the true pilot is one who studies the seasons, wind, seas, and stars. They deny that there is any science relevant to sailing a ship. And if there were a true pilot aboard a ship with these sailors, they would consider him a useless, babbling, stargazer.

For a philosopher, this story ends very poorly. One is left to envision the true pilot at sea, knowing that the ship is more likely to wreck on a reef or sail into a deadly storm than not, but entirely unable to influence the mob of sailors who control the situation. The true pilot has all of the authority and none of the influence or power. Of course, given the shipowner's slight understanding of navigation, there is always the chance that he will side with the true pilot and that the two of them will be able to take the ship in hand; but such speculation goes far beyond Socrates' story, and seems to be a fairly desperate attempt to justify optimism.

The true pilot allegory illustrates one way to describe philosophy's predicament. It is an extension of the classic set of questions implied by philosophy's dependence on and tension with politics. This predicament is no less real than it was in fifth-century Athens, and it may actually be attenuated by the tendency of technology to encourage the opinion that there is little worth knowing that is not immediately useful or cannot be learned quickly. The philosopher, attuned to the way things are, seems detached and irrelevant to those who will not or cannot see their lives exist within a larger context that is more or less intelligible.

Philosophical inquiries, or at least inquiries that present themselves as philosophical, are not entirely without influence in contemporary culture. Most of what falls into this category, however, is pseudo-scientific or mystical. We have all, I'm sure, seen what has become of the philosophy sections of most large-market bookstores. The philosophy section takes up a lot of space, but is filled with New Age mysticism and scattershot collections of Eastern thought, with, occasionally but not always, a much smaller collection of works in the Western tradition. Self-help books used to be regularly shelved with philosophy, too; but, of course, they now demand their own large section.

There are many things to say about the bookstore phenomenon, but I think the foundational issue is that of the plausibility of our ability to engage an intelligible cosmos. The sort of works that call themselves philosophy and that have influence in contemporary culture claim meaning and intelligibility in the world, but quickly retreat to mysticism or agnosticism when it comes to understanding the order of things. It is all phenomenology. No one explains in the Feng Shui manual why hanging something on the west wall of the third room from the front of your house will bring you prosperity, and no one asks for the explanation. But thousands buy the book and hang something up, and the Feng Shui master is venerated for his connection to a cosmic order that neither he nor his students can explain. Mastery of the interactions of inexplicable rules and phenomena is complete mastery.

The primary condition for the possibility of philosophical inquiry is the intelligibility of the cosmos. If the cosmos does not hang together in an intelligible form, then philosophy is a futile, narcissistic hobby. Philosophy is delusion if there is no wisdom to love. On this point, popular contemporary philosophy does fine. Few seem interested, as they were a few years ago, to read philosophers who deny that there is an order to the way things are. Those who read philosophy or anything that tries to have the appearance of philosophy are drawn to the possibility of intelligibility; even the thinnest works now try to satisfy that condition. But it is not enough for a philosophical work to imply the existence of order. Philosophy also requires someone or something capable of appre-hending that order, an intelligence fit to comport itself to the intelligibility of the way things are. The cosmos may be structured perfectly and the whole may exist according to a perfect economy of causes, but if there is no intelligence capable of recognizing that state of affairs, there is no philosophy.

Ancient Greek tragedy explores the boundaries of this view of the human condition by setting exemplary human characters in imaginary universes in which even heroic intelligence is insufficient to make sense of the human condition. The fate of Oedipus eludes understanding. We can, as Oedipus did in the end, see the facts of his condition, but we can tell no story to justify them or even to contextualize them into an account of an intelligible cosmos that could explain them. The cosmos of Sophocles' Oedipus[15] is ordered, but its order is not intelligible to men. In fact, one of the essential characteristics of the Sophoclean cosmos that indicates its unintelligibility is the apparently totalizing determinations on human life. Not only is there order to the universe, there appears to be no detail that escapes necessity. There is no contingency in Oedipus's cosmos.

The intelligibility of the cosmos (and, by extension, philosophy) does not require a totalizing order. One must not affirm pop-philosophical chaos theory and defend the intrinsic connection of all particularity to defend the intelligibility of the whole. Many readers of Aristotle fallaciously reject his notions of final causality because of this mistake. Things have ends for Aristotle, but the ends of some things frustrate the ends of others and there is no necessary cosmic account to justify those frustrations. The acorn could grow into an oak, to use the familiar example; but it could also be eaten by the squirrel. Contingency effects the acorn's fate, but so does necessity. The acorn will not grow into an elm or a cow or a rock.[16]

The story of Oedipus calls into question, perhaps more than anything else, the contingency of human existence. Nothing of significance happened by chance in Oedipus's life. Everything came together fatefully. Thus, he was not only at the whim of cosmic necessity, but in his apparent acts of freedom he was complicit in it. Oedipus illuminates the dark possibilities of an ordered cosmos. While many fear the meaninglessness of life, Oedipus spurs an opposite terror. What if every act of our life has a meaning, and what if that meaning is entirely unintelligible to us?

In Oedipus's cosmos, philosophy is not possible. Tiresias embodies the height of wisdom, and it is not high enough. Blind and old, Tiresias can see a particular aspect of what is, and he can be confident that he has seen it accurately. And, unlike Cassandra, he can often garner the confidence of others in the truth of his vision. Tiresias lacks the ability to do anything about what he knows, though. More importantly, he can never see the way in which his mystical visions resemble or relate to the everyday world he engages blindly.

Tiresias is a scientist. Please forgive the extremity of this comparison—I believe that today's clearheaded scientists do good and important work. After all, Tiresias, if not enviable, is one of the most admirable characters Sophocles gives us. But Tiresias, like even the best scientist, sees only as small part of things. The best scientists see that small part of the cosmos very clearly, having immersed themselves in its vicissitudes with vigor and commitment. But even the most philosophical scientist can say very little about the way in which their narrow field of expertise affects the whole of the way things are. Each connection is another field, and another research plan—another life.

Scientists also, like Tiresias, have little to say about why things are as they know them to be. Tiresias can appeal to the Olympian gods as an intermediary layer of causation, but such accounts are very thin and no one in the play seems to think they are very illuminating. Likewise, a scientist qua scientist has little to say about why phenomena interact as they do. The word, *why* may be used regularly, but it only refers to apparently constant conjunctions between things—on this point, contemporary science is straightforwardly Humean.

Sophocles divides the necessary conditions for philosophical inquiry between Oedipus and Tiresias and then sets them in a world where their excellence cannot find satisfaction. They are like the ship owner and stargazer in the allegory of the true pilot from the *Republic*. Tiresias reads the stars and Oedipus runs the show. If the forces to be ordered by their spiritual and political powers had been mortal, they would surely have prospered. In fact, Sophocles makes clear that they had prospered until the gods sent famine down to punish the murderer of the king. If Oedipus and Tiresias only had to control drunken sailors, they would have had no problem. Instead, the chaos of their situation derived from apparently unruly gods, whom they had no chance of controlling or even negotiating with. Plato's true pilot embodies the authority of philosophy without any influence or power. Sophocles' Oedipus and Tiresias embody together the height of human influence and authority in a cosmos that renders human excellence powerless in the face of a totalizing, unintelligible fate.

As a reader of, I can find very few sanguine things to say about the future of philosophy. Like him I cannot smile at what he saw in the patterns of culture. It seems to me that real philosophical insight is rare and that finding influence and authority attending philosophy such that culture might be conditioned by philosophical truths is far rarer and, perhaps, only possible in a different sort of age. Rather, culture moves steadily toward barbarism. Globalization and technology contribute to the movement of culture farther and farther from its capacity to foster confidence in human wisdom. And, as the literary examples suggest, when human wisdom is not trusted or respected, life moves farther and farther from endeavors that can illuminate human meaning. Yet Ellul's defense of the freedom of individuals to transcend cultural influence suggests a cause for hope. Indeed, since at least some form of philosophical inquiry still seems possible, perhaps there still is hope for insight.

Notes

1. Jacques Ellul, *The Technological Society*, trans. John Wilkinson (New York: Vintage Books, 1964).

2. Ibid., xxviii.

3. Ibid., xxx.

4. Ibid., xxxiii.

5. Ibid., xxxiii.

6. Ibid., xxxiii.

7. Ibid., 428.

8. Ibid., 432.

9. Ibid., 432–33.

10. Ibid., 433.

11. Ibid., 434.

12. Ibid., 435.

13. William Pollack, *Real Boys* (New York: Owl Books, 1999).

14. Christina Hoff Sommers, *The War Against Boys: How Misguided Feminism Is Harming Our Young Men* (New York: Simon and Schuster, 2000).

15. Sophocles, *Oedipus Rex*, in *Three Theban Plays*, trans. Robert Fagles (New York: Penguin, 2000).

16. Aristotle, *Metaphysics*, trans. Hippocrates Apostle (Grinnell, Iowa: Peripatetic Press, 1966).

13

Persons in a
Technological Universe

Donald Phillip Verene

Certain patterns of thought are possible only under certain social conditions. And certain social conditions are prompted only through the presence of certain ideas. Forms of thought and forms of society go hand in hand. Is it possible that the philosophical conception of person and the philosophical conception of truth that underlies modernity—that underlies modern technological life—have a common origin? I think they do. My purpose is to explore this question and to inquire whether the idea of person and the idea of the technological (what in French is called *la technique* and in German *die Technik*) arise together in the modern mind and modern life.

The source for the philosophy of persons and the conception of truth that is the intellectual basis for the ensemble of means that direct modern life is Descartes. This may be surprising to some who look at Descartes's philosophy in a very traditional way. It is not my intention to interpret Descartes as a figure in the history of philosophy, to consider the many sides of his thought, some of which seem quite un-Cartesian, such as his invitation to the reader to conceive his philosophy as an *histoire* or *fable* or approach it as a *roman*.[1] Descartes is certainly the father of modernity, but he is not the cause of the modern world. This world came about through many forces both intellectual and social, but Descartes's thought contains a formulation of the basic principles of this world.

Descartes's *Discourse on Method* contains the conception of truth that is the basis of technological society. What is true and what is real depend upon method and method is understood as a step-by-step process. The first rule of Descartes's well-known method is to begin with the resolution that nothing will be accepted that is not evident and beyond doubt. One begins with something already established with the resolution to establish something more. This is done by dividing difficulties into parts and working from simplest to complex by degrees, as designated by the two middle rules. The final rule is to review what has been made evident to make certain nothing has been omitted.

The most sustained, and my view the most important, attempt to understand the technological basis of modern society as a whole is the work of the French thinker Jacques Ellul. Ellul is controversial among philosophers of technology because they do not understand him. They desire to be specialists working with particular problems concerned with the use of technologies, assessing the effects of such technologies on society, and advocating solutions to the problems they find. Ellul's work forces us and such philosophical specialists to consider the connection between modern society and technology as a whole, to consider *technique* as the motivating and dominating force of all spheres of modern life. I have in mind Ellul's early, comprehensive work, *La technique ou l'enjeu du siècle* (1954) (translated title: *The Technological Society*) and his more recent sequel, *Le système technicien* (1977) *(The Technological System)*, but he is author of many books.[2] In this later work Ellul states that *la seul question philosophique* for our time is the meaning of *technique*.

Put into my own words but reflecting on what Ellul in his 1954 book calls the "technical phenomenon" *(le phénomene technique),* the logic of the ordering of modern life can be reduced to two questions. When anything new is invented or developed or any new state of affairs is realized, two questions are asked, and they are immediately asked: (1) can this be in any way improved? and (2) can it or any aspect of it be applied to areas of activity beyond that in which it has arisen? In this way we live always in the presence of the cult of the new, we exist in a world of "new development." The world of technological order and modernity is a world of everyday wonders. Technique is not something restricted to the working of materials or to producing goods. Technique is the common denominator of our ways of doing business, organizing people, treating the sick, entertaining people, accomplishing work, educating people, producing the written word and the spoken word. To take two initial insights from Ellul and assert them together—the average person is fascinated by performance, and the suggestion box is everywhere.

One is invited not only to eat in a restaurant, one is also asked to improve the service by filling out the card on the table. "Let us know how well we've done." Students are asked not only to study a course but to improve the teaching by completing the evaluation form at the end. We all want to do better and we are all fascinated by performance. The technological person lives in a world of action in which there are no natural beginnings and no natural ends. The television soap opera is the ideal model of technological life. Its action has been going on in many episodes before we began to follow it and it will continue without end after we have ceased to follow its course. The action of the soap opera is Hegel's *schlechte Unendlichkeit,* his "bad infinity"—action that goes on indefinitely with no beginning and no end. Techniques suitable to the educational counseling of students are suitable for managing inmates in prisons, and with modification are also suitable for managing the dying in hospitals and for bracing up herpes sufferers. Techniques used for entertainment through video games are the same found in simulations used in military and business problems. At every turn the individual encounters the future and realizes he must keep up. To repeat Ellul, in no period of history has so much been expected of the individual.

Behind the expanse of technological life is the affirmation of the reality and truth of method. Nothing is real and nothing is true unless it can be formed as a procedure, a step-by-step ordering in which, as Descartes saw, the difficulties can be broken down into parts and we can proceed by degrees from simplest to complex. In doing so there is the drive to completeness and perfection, to interlock techniques and to eliminate error. Technology exists as a system—*un tout organisé.* Technological advance is not the attempt to do everything by one method but the attempt to do anything that is done by method, so that our quest for certainty is continually fulfilled before our eyes. What Descartes could only conceive as an intellectual method for the production of truth, we experience as an actual way of conducting human affairs.

The modern philosophical conception of the person is rooted in Descartes's *Meditations on First Philosophy,* in his famous assertion, I am, I exist. Descartes's philosophy is a dividing line between medieval and modern metaphysics because of his affirmation of the primacy and indubitability of personal experience and his identification of this with mental substance. Descartes's metaphysics of the I is not as such a philosophy of the person, but it is a fundamental precursor of the modern doctrine of person in the way Descartes's epistemology of method is the precursor of the modern mentality of the technical. Descartes as the father of modernity formulates both of these ideas that become governing ideas of modern life.

Descartes's assertion of the existence of the I through his procedure of doubt is not as modern and novel as it might seem. The Italian philosopher Giambattista Vico, in one of his early Latin works, *De antiquissima Italorum sapientia* (1710) (*On the Ancient Wisdom of the Italians*), points out that Descartes's first truth is already present in both Cicero and the *Amphitryon* of Plautus (255–184 B.C.). In several remarkable passages in his work on *The Academics* Cicero suggests doubting the truth of perception because of the experiences of sleeping man and madmen, and he describes the Stoic as having recourse to the idea of a God that can make those things probable that appear to be false and vice versa.[3] This principle is very like Descartes's so-called *malin génie*. More remarkable is the proof of the existence of the I in Plautus's play *Amphitryon*.[4]

In the play, Amphitryon is away at war, accompanied by his slave Sosia. Jupiter assumes the form of Amphitryon in order to sleep with Amphitryon's wife. Mercury accompanies Jupiter in the form of Sosia. While the gods are enjoying the pleasures of his wife and household, the real Amphitryon and Sosia return, only to be viewed as impostors. Moreover, since they find their perfect doubles already present, and going about matters as they themselves, they come to doubt the truth and reality of their own existence. But Sosia remarkably concludes:

> Indeed, when I look at him and recognize my shape, exactly as I have often seen it in the mirror, he resembles me overmuch. He is wearing the same hat and clothes; entirely alike are the calves of his legs, his feet, his stature, the style of his hair, his eyes, nose, teeth, lips, cheeks, chin, beard, neck—in short, everything. If his back is covered with scars, that absolutely completes the resemblance. But, since I think, upon my word, I am and have always been *(Sed quom cogito, equidem certo sum ac semper fui)*.[5]

Because he thinks, Sosia knows he is and has always been. Even the clever deception of the gods cannot deceive him of this truth.

In Plautus's play we have Descartes before Descartes. But let me return from this digression to Descartes as the modern founder of the metaphysics of persons. Certainly in Descartes there is not a philosophy of the person as is to be found in Brightman, process philosophy, or process theology, and even in work on personal identity in analytic philosophy. But Descartes's philosophy, like none before it, asserts the I as the fundamental point of reality, the Archimedian point around which all else metaphysically and epistemologically turns. This endorsement of the primacy of the thinking I is crucial to the technological form of society.

The I is the agent that works the method. Working the method allows the I to master its object—to turn what is inevident and indistinct into what is evident and clear. Truth can be brought out of the indistinct object by the knowing I. But the I understood as thought and thought understood as method immediately creates a condition of *Angst* for the human being, because in this situation the possibility of Socratic self-knowledge is removed.

The I as worker of method has no access to itself. It is the thinking substance that discovers the truth of the object insofar as the object can be thought by the I's method of analysis. But the I's access to itself is restricted to the thought of itself as object, and as object subject to its method of thinking truth. This *Angst* of the I being cut off from itself is displaced in Descartes's *Meditations* and appears as his concern to prove God's existence and nature as independent from the *cogito* through the ontological proof of the Fifth Meditation, and in his concern with the reality of the external world in the Sixth. These are displacements of the I's real fear of its inability to go within itself by its method of truth. In the technological universe this *Angst* issues forth as the deep awareness that technology is not human, namely, that there is no technique of self-knowledge. The technological universe is populated with thinking I's; the *cogito* is everywhere. But these are Eliot's "hollow men." Thought is the working of method, the ordering of the object through technique. Technology provides the human being as thinking I with everything it desires. The I achieves mastery. But on the other hand this achievement of all desires results in nothing. The desires make up the hollow shell that is the modern person.

Thought conceived as method puts self-knowledge and virtue systematically out of reach. The reaction to this *Angst* is to affirm the reality of the person. There is the feeling, quite justified, that the person may slip away and we will be left with the hollow stares and trivial banter and uninteresting language of the technician. Having metaphysically placed all reality in the I, the inner essence of the person, we realize the person itself is most at risk. Where everywhere in the technological society the individual person has no power to determine the course of existence, there is everywhere constant talk of the meaning and value of the individual and the person. Talk of the reality and value of the individual dominates TV talk shows, psychological discussions, and educational theory. All is continually brought back to the blank receptacle of the individual or the person.

Technology is basically about desire. The technical impulse is rooted in the larger phenomenon of human desire in the sense that Hegel has analyzed desire *(Begierde)* in his famous discussion of Mastery and

Servitude *(Herrschaft und Knechtschaft)* in the *Phenomenology of Spirit*. The achievement of the reality of the self, to be actually a self, is, as Hegel formulates it, a life and death struggle. This requires that conditions exist such that the individual can risk his life in order to come to know that he is truly alive as a particular, living self. In the technological universe the self is cut off from the conditions of genuine struggle. Because of the power of technique to master what is present as object, the individual can satisfy his desires. Desire as a fundamental phenomenon of human existence can never lead to the life and death struggle of the self with its world, with the other. Desire can never take root in the technologically equipped person because he always has the power to adjust the object partially to his desire.[6]

Hegel says: "The individual who has not risked his life may well be recognized as a *person [als Person]*, but he has not attained to the truth of this recognition as an independent self-consciousness."[7] The I as agent of technological activity risks becoming a person in the sense of having only generic existence. To live as a person in Hegel's sense is to exist as an abstraction. One wishes to be recognized as something more than a person. One desires to be recognized at least as a person but one desires to be something more than that. Merleau-Ponty, on being asked to describe a person, replied he was unable to do so unless he knew what sex it was, whether it was a man or a woman. The problem in technological life is that all humans are immediately persons. And because all human beings are persons no one is a person. It is like Hegel's insight at the beginning of the *Science of Logic*. Pure being without any determination is the same as pure nothingness. To think being without any determination—as something true of anything—is just empty thinking.

Descartes could only intellectually formulate what it would be like to be an "I think" in person—the master of the object who has no access to the inner act of self-consciousness because his being is not connected to desire in such a manner as to take it into Hegel's life and death struggle. The Socratic project is closed to the modern person partly because the language of the Socratic *agorà* is closed down by the technological mastery of the object.

Philosophical interest in the person and the technological form of society must face two problems, presented by Giambattista Vico as among the first axioms of his new science. These are what Vico calls the *boria delle nazioni,* the conceit or arrogance of nations and the *boria de' dotti,* the conceit or arrogance of scholars. The conceit of nations, as Vico describes it, is that a nation claims that it "before all other nations invented the comforts of human life and that its remembered history goes back to the very beginning of the world."[8] The technological form of society acts as if it were the only desirable form of life and presents itself as if all human history has always exhibited a tendency toward technique. Thus, our age appears to be a fulfillment of all the hopes of past ages. This is the common view that humans from the beginning have made tools and discovered better ways for doing things. It is true that wherever there is evidence of human society there is evidence of tool use. There are not human societies without tools and techniques. But in such societies there is not technical consciousness involving the two questions that I described above. I agree with Ellul that Western society takes *technological shape* beginning in the 1750s, with the use of machines as means for production, and that the machine is only the first form of meaningful technique. Technique appears originally in the form of the machine, but is not limited in its being or nature to it.

The conceit of scholars is the danger encountered by attention to the idea of persons because it seems such a natural idea. It is a natural idea to the modern mind. Vico warns that the conceit of scholars is that scholars "will have it that what they know is as old as the world."[9] What is really a modern idea—persons—seems to have always been the right thing to have. The modern term *person* in European language derives from the Latin term *persona,* preserved most directly in modern Italian, which uses precisely the same word. In the modern pursuit of person, the ancient sense has been put aside. By way of conclusion I would like to recall this original sense and say a few words on its behalf.

Persona is a mask used by a player, a character or personage acted *(dramatis persona)*. The term also has an early legal sense and the Christian sense of a person of the Trinity. But it is the sense as mask in which I am most interested. Descartes's I can wear no mask. His thinking being can doubt, understand, conceive, affirm, deny, will, reject, imagine, and perceive; but it cannot mask. The cognitive I is condemned to seek the truth, to seek right reasoning by method in the sciences. Such an I cannot enter the world as a theater. It cannot participate in what in the view of Erasmus is the "theater of mortals" or what Sebastian Brant called *Narrenschiff,* the Ship of Fools.[10] There is no double life to the Cartesian *cogito.* It is always the same, always on the same single-minded quest for the truth. The dramatic is closed to it.

In fact, it is the mask-like character of the human self that disturbs Descartes—its ability to appear one way in sleep, another when gone mad, never to show itself with an indubitable, real face. The perceiver cannot know if the other beings it sees in the street are persons like itself or whether there are merely hats and cloaks covering automatons. Descartes shows the way out of such confusion by showing the essence of the self to be the single I, unmasked behind its masks. The modern concern to get at the true person behind its appearances is evident in the concern for roles as found in Existentialism (one thinks of Sartre) and in the concern of modern psychology, popular and otherwise, with the discussion of roles and the attempt to "find oneself."

The difference between Sosia's formulation of *cogito ergo sum* and that of Descartes is that Sosia is part of a *dramatis personae*. His *cogito* is one of his roles. It is the one bound up with the figure who has whip marks on his back and who has a particular haircut and hat size. Descartes's *cogito* has no such form. It has no whip marks on its back. It has not returned from the wars to find gods who have taken its form enjoying the pleasures of the ladies of the household, the ladies happily in the rapture of the gods.

Persona suggests a life lived through its succession of masks, a life that approaches the human world as a theater of mortals with the fascination of playing roles. *Person* and the achievement of *personhood* suggests metaphysical and moral respectability that hopes to avoid the mask in favor of an account of true selfhood. I do not mean this is a bad aim, as it is based on the need to establish a human nature in the facelessness of the technological universe. But the problem the philosophy of persons confronts is the sense of a particular life lived apart from the search for a Cartesian center, one that can formulate the *cogito* not in general but only in front of Sosia's mirror.

Notes

1. *The Philosophical Works of Descartes,* trans. Elizabeth Haldane and G. R. T. Ross, 2 vols. (Cambridge: Cambridge University Press, 1931), I, 83 and 209.

2. Jacques Ellul, *La technique ou l'enjeu du siècle* (Paris: Colin, 1954); *The Technological Society* (New York: Knopf, 1964); and *Le système technicien* (Paris: Calmann-Levy, 1977); *The Technological System* (New York: Continuum, 1980).

3. Cicero, *Academica priora II*, 15, 17.

4. Plautus, *Amphitryon*, 441–47.

5. Translation is from *Vico: Selected Writings,* ed. and trans. Leon Pompa (Cambridge: Cambridge University Press, 1982), 57.

6. Donald Phillip Verene, "Technological Desire," in *Research in Philosophy and Technology,* vol. 7, ed. Paul T. Durbin (Greenwich, Conn.: JAI Press, 1984), 99–122.

7. G. W. F. Hegel, *Phenomenology of Spirit*, trans. A. V. Miller (Oxford: Oxford University Press, 1977), 114.

8. *The New Science of Giambattista Vico,* trans. Thomas Goddard Bergin and Max Harold Fisch (Ithaca: Cornell University Press, 1968), par. 125.

9. Ibid., par. 127.

10. Donald Phillip Verene, "Technology and the Ship of Fools," in *Research in Philosophy and Technology,* vol. 5, ed. Paul T. Durbin (Greenwich, Conn.: JAI Press, 1982), 281–98.

Contributors

IAN ANGUS is a Professor of Humanities at Simon Fraser University. He is the author of *(Dis)figurations: Discourse/Critique/Ethics* (Verso, 2000), *Primal Scenes of Communication: Communication, Consumerism, Social Movements* (State University of New York Press, 2000), *A Border Within: National Identity, Cultural Plurality, and Wilderness* (McGill-Queen's Press, 1997; reprinted 1998), and *George Grant's Platonic Rejoinder to Heidegger: Contemporary Political Philosophy and the Question of Technology* (Edwin Mellen Press, 1988). He is the editor of *Cultural Politics in Contemporary America,* co-edited with Sut Jhally (Routledge, 1989) and has published widely on technology, communications, and Canada.

BERNARDO ATTIAS is currently an Associate Professor in the Department of Communication Studies at CSUN, where he has taught since 1994. His research focus emphasizes cultural approaches to communication studies as well as communication-centered approaches to cultural studies. The emphasis of much of his work is on the political economy of mass mediated events. He has written on media coverage of the war in the Gulf, on the politics of psychoanalysis, on the rhetoric and politics of hip-hop culture, and on the drug war, and he has introduced an interdisciplinary notion of "internal third worlds" useful for the analysis of urban geopolitical phenomena such as the shifting demographics of Los Angeles.

DARIN BARNEY is a native of Vancouver and was educated at Simon Fraser University and the University of Toronto, where he was trained in political theory and received a Ph.D. in 1998. He is the author of *Prometheus Wired: The Hope for Democracy in the Age of Network Technology* (University of British Columbia Press/University of Chicago Press/ University of New South Wales Press, 2000). He is currently Assistant

Professor of History and Politics at the University of New Brunswick in Saint John, Canada, where he also directs the Information and Communication Studies Program.

TOM DARBY is a Professor of Political Science at Carleton University in Ottawa, Canada. He is the author of *The Feast: Meditations on Politics and Time* (2nd edition, 1990), "On Who Has the Right to Rule the Planet" (in *Kritika and Kontext*, 1999), "Power and Wisdom: Technology, Christianity, and the Universal and Homogenous State" (in *After History*, T. Burns, ed., 1994), the editor of *Sojourns in the New World: Reflections on Our Technology* (1986), and co-editor of *Nietzsche and the Rhetoric of Nihilism* (with Egyed and Jones, 1988).

ANDREW FEENBERG is a Professor of Philosophy at San Diego State University. He is the author of *When Poetry Ruled the Streets: The French May Events of 1968, Questioning Technology, Alternative Modernity, Critical Theory of Technology, Lukács, Marx, and the Sources of Critical Theory,* and co-editor *of Technology and the Politics of Knowledge*. While working at the Western Behavioral Sciences Institute in the early 1980s, Dr. Feenberg participated in the development of the first educational program delivered over a computer network. He has pursued his studies of the philosophy of technology and educational applications of computers with support from the National Science Foundation, The U.S. Department of Education, and the Digital Equipment Corporation. He lives in La Jolla, California.

GILBERT GERMAIN is an Assistant Professor at the University of Prince Edward Island. Author of *A Discourse on Disenchantment: Reflections on Politics & Technology* (Series: SUNY Series in Political Theory: Contemporary Issues) published in January 1993 by State University of New York Press.

TRISH GLAZEBROOK received her Ph.D. from the University of Toronto in 1994, and has taught at the University of Toronto, Colgate University, and Syracuse University. A board director of the International Association of Environmental Philosophers, she is the author of *Heidegger's Philosophy of Science,* the editor of a forthcoming collection of essays on Heidegger's critique of science, and publishes on Heidegger, ancient and modern science, environmentalism, technology, and feminism.

HORST HUTTER received his Ph.D. from Stanford University and is Associate Professor at Concordia University. He is the author of "With the 'Nightwatchman of Greek Philosophy': Nietzsche's Way to Cynicism," in Bela Egyed and Tom Darby, eds., *Nietzsche and the Rhetoric of Nihilism*, 1989; "Philosophy as Self-Transformation," in *Historical Reflections/Reflexions Historiques* 16:2–3 (1989): 171–98; *Politics as Friendship* (Waterloo: Wilfred Laurier University Press, 1978).

DON IHDE is a professor at the State University of New York at Stony Brook. He is a leading philosopher of technology and author of *Philosophy of Technology: An Introduction, Technology and the Lifeworld: From Garden to Earth*, and the seminal *Technics and Praxis*.

ARTHUR MELZER is Associate Professor of Political Science at Michigan State University. He is the author of *The Natural Goodness of Man: On the System of Rousseau's Thought* (1990). He is a director of the Symposium on Science, Reason, and Modern Democracy.

WALLER R. NEWELL is a Professor of Political Science and Philosophy and co-director of the Centre for Liberal Education and Public Affairs at Carleton University in Ottawa, Canada. His publications include *The Art of Manly Virtue, An Anthology with Introduction and Commentary* (Regan Books/Harper Collins), *Ruling Passion: The Erotics of Statecraft in Platonic Political Philosophy* (Rowman and Littlefield), *Bankrupt Education: The Decline of Liberal Education in Canada* (with Peter C. Emberley) (University of Toronto Press, 1994). His teaching and scholarship are focused on the history of political philosophy. His specializations include classical political philosophy (including Plato, Aristotle, and Xenophon) and German Idealism with its ramifications for contemporary phenomenology, critical theory, and postmodernism (including Hegel, Nietzsche, and Heidegger).

CHARLOTTE THOMAS received her Ph.D. From Emory University in 1996 and is Assistant Professor of Philosophy and Chair of the Department of Interdisciplinary Studies at Mercer University in Macon, Georgia.

DONALD PHILLIP VERENE is the Charles Howard Chandler Professor of Metaphysics and Moral Philosophy, Emory University. He is also the Director of the Institute for Vico Studies, Emory University, and past Chair of the Department of Philosophy, Emory University (1982–1988). He is the author of *Philosophy and the Return to Self-Knowledge* (Yale University Press, 1997); *The New Art of Autobiography* (Oxford: Claredon, 1991), *Hegel's Recollection* (State University of New York Press, 1985), *Vico's Science of Imagination* (Cornell University Press, 1981) and the editor of *Symbol Myth and Culture: Essays and Lectures of Ernst Cassirer 1935–45* (Yale University Press, 1979), *Vico and Joyce* (State University of New York Press, 1987).

Index